Cars and coachbuilding

One hundred years of road vehicle development

Cars and coachbuilding

One hundred years of road vehicle development

GEORGE OLIVER

Sotheby Parke Bernet

Published in association with the Institute of British Carriage and Automobile Manufacturers to celebrate its centenary, 1881–1981

First published 1981 in association with IBCAM for
Sotheby Parke Bernet Publications by
Philip Wilson Publishers Ltd
Russell Chambers, Covent Garden
London WC2E 8AA

and

Sotheby Parke Bernet Publications
Biblio Distribution Centre
81 Adams Drive, Totowa
New Jersey 07512, USA

ISBN 0 85667 105 3

Designed by Gillian Greenwood

Filmset and printed in Great Britain by
BAS Printers Limited, Over Wallop, Hampshire
and bound by Mansell (Bookbinders) Limited,
Witham, Essex

Contents

Foreword

This record of one hundred years of coachbuilding is a centenary tribute to the Institute of British Carriage and Automobile Manufacturers. I congratulate the author, George Oliver, and the publishers, for producing what is probably the most significant and impressive historical account of the transport industry and the Institute's involvement over the past hundred years.

The Institute was founded in 1881 by prominent coachbuilders of the day to safeguard the interests of those engaged in the coachbuilding industry and to promote the advancement of technical education. In recent years the development of new methods and the use of a wider range of materials for the construction and finish of motor bodywork have provided scope for the extension of the usefulness of the Institute as a centre for the exchange of ideas and the publication of the results of individual experience which are made available for the benefit of the whole industry. For this purpose, activities such as lecture and discussion meetings, film evenings and works visits are arranged by the six local centres which now cover the whole of England, Scotland and Wales.

Special features of the Institute's activities are the arrangement of competitions for the purpose of improving body design. Much the most important of these are the coachwork competitions held in conjunction with the British International Motor Show, for which the great majority of cars are entered. The National Bodycrafts Competition for Apprentices and the National Drawing Competition for Apprentices are arranged with the active assistance of the local centre committees.

I express my appreciation to the author for the presentation and painstaking research undertaken by him and to the members of the Institute who assisted him in producing a book which provides most interesting reading and which, furthermore, will become a reference for present and future generations involved in the profession of coachbuilding.

Peter B. Cheston, FIBCM
President IBCAM

Preface

It is easier to make a survey of a major industry if one is not a part of it oneself; from the outside one can more readily take the long view: certainly it is possible to see things with greater objectivity. This book is a 'looking in', as it were, by someone who has been looking on for the past fifty years, precisely half the life-span of the Institute of British Carriage and Automobile Manufacturers, whose centenary in 1981 it celebrates. In no way does the book try to be either a detailed history of the Institute itself or of the great industry with which that organisation has been so closely connected for the past century, the length of the period covered and the vastness of the subject making such a task a practical impossibility.

Looking in is all very well, but it helps the outsider if there are willing friends on the other side to open doors and provide access to richly stocked memories. Here the writer has been helped enormously by the present President of the Institute, Peter Cheston, the Secretary, Peter Hawkes, and by four influential members, whose combined store of knowledge and practical experience have provided the best possible source of first-hand information that a transport historian could hope for: Leslie Ireland, George Moseley, Eric Neale and Ken Osborne, from whom a great deal has been learned. He has also had the privilege of regular consultation with his friend Michael Worthington-Williams, whose advice, drawn from an Ali-Baba-like storehouse of automotive knowledge, has been received with gratitude and acted upon with alacrity. What follows is a sampling of the industry's achievements over the most exciting period of its history.

G. Oliver, 1980

1 *Between the wars in Britain, and long before the present-day revival of interest, a few enthusiasts used lovingly-cherished horse-drawn vehicles on suitable occasions. One of them was the circus owner, Captain Bertram Mills, who ran this coach, 'The Old Times', between the Berkeley Hotel, London, and Windsor. The exposed situation of the outside passengers is clearly to be seen. The picture was taken* circa *1930, in Sheen.*

I Before the motor-car

By 1881, the railway had become the most important and popular form of medium- and long-distance transport in Britain. In little more than forty years the social and economic life of large areas of the British mainland had been gradually transformed by the inexorable growth of systems of communication that telescoped time and distance. It was no longer necessary to suffer eight days of potential danger and actual physical discomfort, of a kind scarcely to be imagined nowadays, to travel west, from London to Exeter, for example; or to set aside at least twelve uncertain winter days to chance the 400 mile journey north, to Edinburgh. It was on these roads to the north and west of the capital, incidentally, that the best coaches were used: sturdiness was an essential feature—well-exemplified in the mail coach [1].

The hazards and practical difficulties of road travel of any length were glossed over by many writers, who had personal experience to draw upon but chose instead to indulge romantic fantasies as far from truth as the traditional Christmas card notion of coach and horses swishing through snow unhindered, with beaming crew and passengers—the latter crammed inside their conveyance or precariously perched outside, upon its roof.

Weather was by far the greatest of the many problems facing the road traveller in the last century, or before. It was the cause of others, too, the worst of which was its ruinous effect on roads themselves. Conditions akin to those experienced nowadays by drivers in cross-country trials or off-road exploration were commonplace in periods of rain or snow. At other times dust was the greatest difficulty. There was, of course, no such thing as the Land-Rover or power-winch: no motoring organisations with radio control of twenty-four-hour rescue and recovery services; neither they nor the helpful, hovering helicopter had yet been invented. The coach that suffered structural breakage or other kind of mishap far from town or village could be stranded for a long time before help appeared. And even when all seemed well, when vehicle and horses matched to near-perfection, when it was neither too hot nor too cold, when highwaymen were absent from the scene (not that they were often on it, in fact), and when travelling companions were of a generally agreeable kind, there still remained the chance of sudden weather-worsening, and apprehension as to the quality of services to be encountered at the next scheduled stop. Variable standards of main-road catering in Britain are not new.

There were compensations. Travel by horse-drawn vehicle was a great deal swifter and less fatiguing than making the same journey on foot or horse-litter, for example. Under ideal conditions the exhilaration of movement behind units of motive power as fine as healthy, vigorous, willing and well-teamed horses had strong attractions; there was, too, a sense of achievement at journey's end shared by all involved, whether in an active or a passive way. Above all travel of this sort brought those concerned into comprehensive contact with the countryside through which they passed—an experience only to be enjoyed now by horse-riders, cyclists, walkers or the occupants of open motor cars. There was, moreover, an understandable feeling of satisfaction in ownership of a really fine vehicle [II]. The excellent appearance of most horse-drawn vehicles was another favourable factor. Apart from state coaches and some other ceremonial carriages that were heavily encrusted all over, simplicity of shape and subtlety of proportions were leading characteristics, common to almost all types and sizes, from grandest travelling coach to tiniest pony cart. Traditional, internationally recognised British skills in the design and manufacture of road-passenger vehicles had been continuously refined and improved over centuries, quality of appearance being matched by high standards of craftsmanship. Long after the ascendancy of the motor had been completed, a number of enthusiasts continued to use specially cared-for examples [1].

In some mysterious manner the Victorian addiction to decorative excess, shown to the world in all its fulsome horror at the Great Exhibition of 1851, did not affect the craft of coachbuilding to any noticeable degree. This was a remarkable state of affairs for the time, particularly in an industry so closely linked to fashion at its higher levels. The fact remains, however, that the timeless look of the best nineteenth-century vehicles was almost untouched by contemporary influences. General form was restrained, but they were in no way dull or austere, excellence of shape being matched by comparable excellence of fit, finish and colour. Those who sought the rococo had to go abroad, (to France, Spain, Portugal, Italy or Austria, for example), to find 'fine ornamental carriage-building' [2].

In other respects, however, the coachbuilding trade could not remain unaffected by contemporary developments. The principal challenge, commercially speaking, came from the railways which, at an extremely early stage in their history, had taken the long- and medium-distance transport of passengers away from the coaching interests. Within a year or so, systems based on the cumulative experience of centuries and raised to high levels of operational efficiency were eliminated. With them went a whole way of life for countless thousands.

The consequences were widely felt, not simply by the coach operators and their numerous employees, but by all involved in servicing and supply; horse-breeders and dealers, for example; hay and feed merchants, saddlers and harness makers, blacksmiths and wheelwrights, stable-keepers, the proprietors and staff of roadside inns and the coachbuilding trade itself. The market for the larger kind of vehicle, in particular for mail and stage coaches, vanished, more or less completely. At a later date, the spread of suburban rail services, particularly in the London area, led to the reduction in the use of privately owned vehicles, noted by Mr G. N. Hooper in 'Modern Carriages', his richly interesting contribution to the Badminton Library

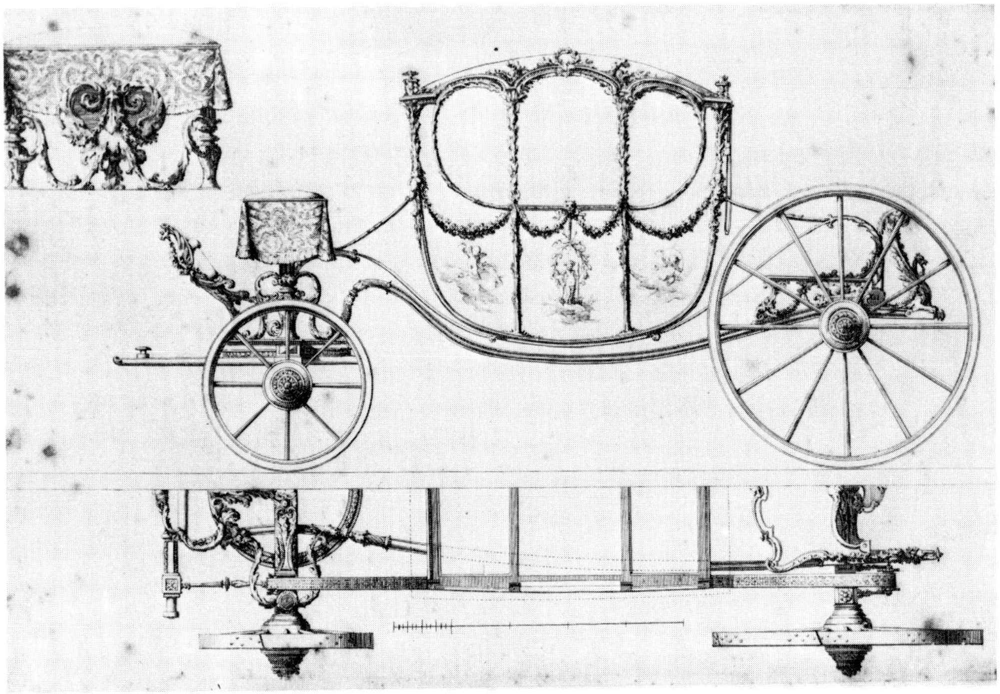

2 *Rococo on wheels—possibly French. The undated original drawing is captioned* A State Coach on two perches, with long leather traces, *and shows a vehicle of basically simple shape combining great elegance with power of appearance. Applied ornament of this kind was more common on the Continent than in Britain.*

volume, *Driving*, first published in 1889. After pointing out that 'The rough and heavy travelling and other carriages that preceded 1830 fell out of use as the new and better roads of McAdam were made', he referred to the mail and stage coaches which 'had just reached perfection in design, durability, lightness and handiness, when the introduction of railways literally drove them off the high roads that seemed to have been made for them', and went on to observe that 'the gigs and phaetons, kept in large numbers by the London bankers and merchants to drive to and from their suburban houses, were driven off the roads by omnibuses, tramcars, and suburban railways'. As these public services were more notable for discomfort than anything else in their early stages of development, it might be fair to assume that novelty was a principal attraction at first. Later on, all-round improvements made their cheapness and convenience almost irresistible. Nevertheless, many owners of vehicles as handsome and versatile as the Brougham [111] and its variants, went on using them, mainly for town work. The gentleman's omnibus was better suited for country use, having something of the versatility of the present-day estate-car.

By the beginning of the 1880s, the number of horse-drawn vehicles in use in Britain was well over 400,000. In *Coachbuilding Past and Present,** Cecil Robertson gave the number of 'carriages of every description' as 60,000 in 1814 and 435,000 in 1874. In 1834, there were 49,000 four-wheelers and 50,000 two-wheelers in the

* Cecil Robertson, *Coachbuilding Past and Present*, Hooper & Co., 1928.

United Kingdom. Forty years later these totals had risen impressively to 150,000 and 285,000, respectively. Nevertheless, the country's coachbuilders were unhappy about the state of their industry.

On 6 December, 1881, a large and representative group from London and the provinces met in the capital to review the current situation as far as it affected their interests, and to consider what practical advantages there might be in the establishment of a 'combination', or association. At that time their craft had been in a depressed state for many years; their once-prosperous export trade had all but vanished and at home, a growing tendency on the part of buyers to turn to foreign products greatly worried them. The ascendancy of the railways had totally destroyed their former lively business in the larger long-distance coach, although an increasing demand for buses partly compensated for this—a fact that was not referred to. Above all, taxation was thought to be a 'serious obstacle to trade'.

Why this should have been so, it is now difficult to comprehend. At that time the annual tax levied on a two-horse or four-wheeled vehicle, for example, was two guineas, which seems little enough by our standards. It was thought excessive then, however, and the trade had been fighting hard for repeal, for years past, pointing out, quite properly, that coachbuilding and silversmithing were the only manufacturing industries in the country so penalised. Taxation was not new as far as road-vehicle makers and users were concerned. King Charles I had imposed a tax on coachbuilders as early as 1637, but he had also encouraged their activities, which was more than the Puritans had done during the Commonwealth. In 1662, after the Restoration, there were 400 hackney coaches in London, on each of which an annual tax of £5 was paid. In 1790 the rate for a four-wheeled carriage was £8 16s 0d; for a two-wheeler it was £3 17s 0d—but if the latter cost less than £12 new and carried the words 'Taxed Cart', this was reduced to 12s. From 1 January 1870 altered rates came into force, the charge on carriages with four or more wheels and weighing more than 4 cwt being two guineas. Vehicles below that weight were taxed at 15s a year; so, too, were those with less than four wheels weighing above 4 cwt. Each horse or mule was charged at half a guinea per annum. From 1880 revised rates applied. An equivalent of the present-day family saloon car was the Double Brougham, which carried four inside, two outside, and cost anything between 100 and 185 guineas. In relation to this purchase price, and to yearly running costs which in London amounted to about 180 guineas a year, such a rate of tax was not, surely, a punitive one.

Even when road toll charges were taken into account the private owner did not have to pay out too much in taxation, directly or indirectly. For the commercial operator, however, tolls added significantly to his running costs. Morever, he was subject to a multitude of rules and regulations framed, as so often seems to be the case throughout history, by bureaucrats with too much time on their hands and much too little of anything inside their heads. The carriers of goods were most affected by constructional requirements that were frequently ridiculous; such matters as the width of wheels, their distance apart, vehicle weights and the number of horses or other draught-animals being juggled with in different ways, at different times, to little, if any, practical purpose. From the passing of the first Parliamentary Act, in

1663, until the end of road tolls altogether in the mid-1890s, users of the turnpikes had to pay predetermined levies. For example, these were a penny for a horse, sixpence for a coach, eightpence for a cart and a shilling for a wagon in the seventeenth century. On a long day's journey the total could be substantial by contemporary standards, and while surplus revenue was supposed to be used for the improvement of the roads administered by individual Turnpike Trusts, high administrative costs and almost universal misappropriation of tolls received meant that there was seldom much left over for that admirable purpose.

Adventurous, far-sighted individuals were trying to establish steam-driven, road-coach services in the 1830s, but the attitudes and subsequent actions of certain Turnpike Trusts, which had persuaded the not-unwilling Parliament of the day to let them raise toll charges to unrealistic levels, made successful commercial operation quite impossible. As the swift spread of railways began to draw off more and more of the goods and passenger traffic hitherto transported by road, revenue continued to fall, along with standards of highway maintenance. The wonderful work of road improvement, begun by the blind John Metcalf in Yorkshire in the 1760s and carried on with even more success by the Scotsmen, Thomas Telford and John McAdam, in the late eighteenth and early nineteenth centuries, was almost undone.

For thirty years or so, through-traffic was practically non-existent on the principal highways in Britain. The superbly-organised coach services between towns and cities which, at best, set exceptionally high standards of passenger care, turn-out and punctuality, despite all practical difficulties, had gone well before the mid-nineteenth century. As we have already learned, their passing was accompanied by the passing of a way of life intimately linked to the horse, the carriage and the road. Much the same was to happen to those connected with another form of transport, the canal. Its long-boats, their crews, the locks and lock-keepers, landlords of waterside inns and all those concerned with supply, repair and maintenance services were slowly to disappear from a scene as distinctive, varied and colourful, in its way, as that of the road.

Matters of this kind were not the immediate concern of the coachbuilders present at the historic meeting in December 1881. They busied themselves with setting about the establishment of what was shortly to become known as the Institute of British Carriage Manufacturers. Its main aims were to watch over the interests of its members, to promote their prosperity, to provide suitable meeting places, to consider and discuss such matters as taxation, foreign tariffs, rail and sea-freight charges, insurance and so on, to advance their trade and to help with the education of young people entering the industry, either as future employers or employees. The effectiveness of associations formed in other trades had been closely studied and the success of the Syndical Chamber of Coachmakers in Paris and of various American coachmakers' organisations were well-known to many members of the British industry. There was also an awareness of the practical results of technical teaching in France and its favourable influences there, in other European countries and in the United States. As part of the educational process, it was decided that the new organisation should establish an efficient Reference Library and make a collection of drawings and photographs of carriages used in different parts of the world, many of

4

3 5

3 *Such exuberant details as these were designed for the showiest and costliest vehicles; in this instance they are suggestions from France for the 'hind-carriage of the first gilded coach', and the original sheet of drawings was an early acquisition for the IBCAM Library, along with the elegant elevations seen in plates 4 and 5.*

4 *Even when allowance is made for the conventions of the carriage draughtsman (who slimmed down wheels and springs, for example) the fine proportions and elegant appearance of this coupé (a derivative from the English Brougham), designed and built by Henri Binder of Paris, are unmistakable. Binder's firm survived the gradual changeover from carriages to cars and built automobile bodies until 1945 (above, right).*

5 This voiture de commerce, *by Mm Vareval & Vacher, of Paris, dates from 1879 and has a delicacy of look that would make it especially suitable for use by the fashion trade (below, right).*

6 *At regular intervals,* The Coach Builders' and Wheelwrights' Art Journal *published sets of large-scale working drawings of a great variety of horse-drawn vehicles that provided subscribers with all the information necessary for construction. The illustration shows part of a set for a mail coach and apart from the excellence of its design it is notable for the high quality of its draughtsmanship.*

7 *Working drawings were understood by the coachbuilder and his highly-skilled craftsmen. Drawings of the type shown here (and elsewhere) could be understood by the customer, for whom they were created specially. This elevation shows a 'Grand Break de Chasse', an 1879 design by Henri Binder, that provides distinctive appearance with the practical advantages of elevated seating, rather in the manner of the present-day Range-Rover.*

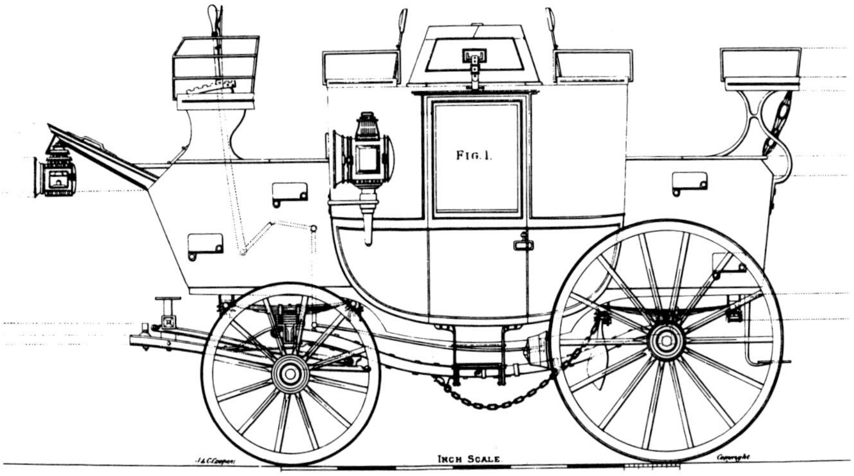

6

7

which appear in this book. Plates 3, 4 and 5, for example, were early additions to the IBCAM Library, from French sources. Moreover, it was agreed that it should promote 'creditable displays' of British carriages at great national and international exhibitions. Even before exhibitions had become popular, there had been a constant interchange of ideas and information between the coachbuilders of different countries; certain vehicle types found favour outside their own national boundaries and became known internationally. Such names as Barouche, Britzska, Berlin, Cabriolet, Dogcart, Landau, Stanhope, Victoria, Vis-à-Vis and others crossed the English Channel and the Atlantic Ocean in both directions. The process continues. It was helped greatly from the middle of last century by the publication (mainly in London, Paris and New York) of a number of specialist books and journals providing practical information on all matters relating to the coachbuilding trade, as well as a regular supply of design ideas, working drawings and, in some cases, colour plates as well. In Britain, and in many other countries where it was eagerly studied, *The Coach Builders' and Wheelwrights' Art Journal* (an enterprise of the resourceful and talented carriage designer and draughtsman, J. Cooper) maintained consistently high standards. Plate 6 shows a detail from a typical Cooper working drawing and 7 is representative of French work of the same period.

Although George Hooper, the first President of the Institute from 1881 until 1890, was prepared to admit that home and international exhibitions played a part in stimulating changes and improvements ('the former much more numerous than the latter') he warned of the possible dangers of exhibiting in other countries. It was not unknown for the inviting country to reserve the best situations for its own manufacturers and to place foreign competitors in distant and unfavourable settings, hard by displays of agricultural implements and vehicles, for example. Gold medals, he wrote,* would not make up for the unfortunate impression thus made on potential customers; and rival manufacturers of the host country were always at hand, to take patterns and measurements neatly and accurately.

His view was, perhaps, a little jaded. Public exhibitions serve a useful function as clearing-houses for new ideas, new methods, new fashions, and the traffic is by no means all one way. Early in the twentieth century, as John Norris recorded in his brief history of the Institute, (published in 1961 by IBCAM), 'the British Coachbuilders set off for Paris to see what the French, who had seized the lead, had to show at the Grand Palais. New, strange names and ideas were collected, copied, absorbed and improved upon. So, within two or three years, the English coachmaker was holding his own with his competitors from France, Germany, Belgium, Italy and America at Olympia. All learned from one another.' Much the same could be said today, with equal truth. The design and manufacture of road vehicles is an international matter.

In one way or another this has been so for centuries. For example, the Cabriolet originated in France but was taken up in Britain and a good deal improved upon by some of the London coachbuilders. Early in the reign of Queen Victoria, Count d'Orsay, 'a man of fashion and some taste in carriages', turned his attention to this type and, in association with Mr Courtney, of Barker & Co., made further improvements. Various slight but significant alterations to its shape and look were made and 'such high-stepping, grand-action horses as made the new and refined carriage look worth twice as much as its predecessor' were selected. An earlier type of carriage, the Berlin, was introduced in Germany, in 1660. It was the creation of Philip de Chiesa, a Piedmontese employed by the Elector of Brandenburg. Italian involvement in the design and construction of horse-drawn vehicles reached back much further; the Roman chariot, or *essedum*, the Etrurian car, the *cisium*, the *sarracum*, *plaustrum* and *carpentum* being but a few of the types developed in that land of miraculous mechanical and aesthetic invention. Closely linked to vehicle development in Italy, was the creation of a precisely planned system of well-engineered roads, which served peaceful as well as military purposes, linking continents as well as cities. It is interesting that the *cisium* (like the chariot, a forerunner of the gig), was so popular in Rome that it was available for hire; that the *carpentum* was reserved for the use of distinguished citizens, at one time, and became notable for the richness (though not necessarily the good taste) of its ornamentation, and that while the *pilentum* was extremely popular for official occasions, the matrons of Rome were only allowed to drive in it at the games or at public festivals. Italian

* *Driving*, Badminton Library, 1889.

8 and 9 *The builders of these goods vehicles thought it important that their products should look well and work well. Although they are sturdily built there is nothing clumsy about their shapes overall, and good lettering and clever use of decorative 'lining' (on wheel spokes, rims, and body panels for example) adds interest to the general effect.*

road vehicles of our own time are distinguished by the basic simplicity of their shape and by an almost uncanny feeling for visually exciting and satisfying forms. Twenty centuries ago, it would appear, their predecessors relied somewhat more on applied ornament, on the reckless use of precious metals, richest silks and other fine materials, paintings and pieces of sculpture. For some time, indeed, the amount of decoration on private carriages was limited by decree, in the Roman sumptuary laws.

During the reign of the Emperor Alexander Severus, these restrictions were lifted; carriage building grew into an important, prosperous activity. Much the same happened after the return of the monarchy to Britain, following the Interregnum of 1649–60. Charles I, as we have seen, encouraged the coachbuilders, and Charles II was even more helpful; in 1677 he founded the Company of Coach and Coach Harness Makers, an organisation that by the terms of its charter had the power to search out and destroy work of poor quality and insist on high standards of craftsmanship. After a life of ten years the Company lost its charter; a year later, however, it was re-granted in the name of the Worshipful Company of Coachmakers and Coach Harness Makers of London. Under its rules no one could trade lawfully as a coachmaker or coach harness-maker within London and Westminster or the country within a twenty mile radius of the capital, unless he was a member. In the 1860s the Company began to provide prizes to encourage the maintenance of high standards of vehicle design and workmanship. At a later date the Institute also became involved, along with the Society of Motor Manufacturers and Traders and the National Federation of Vehicle Trades. For a very long time competitions have played an important part in the educational activities of the Institute, offering, as they do, opportunities for young designers and draughtsmen to try their skills in many new, perhaps unexpected, directions.

At its best the carriage has long been something more than an example of fine craftsmanship; a splendidly balanced combination of craft and art, its varied types gradually developed over centuries, from practical experience, until forms reached peaks of excellence, functionally and aesthetically speaking. In Britain coachbuilding remained a craft-based industry, its members widely spread, in a geographical sense, but close indeed in an almost universal concentration on quality. It is an interesting fact that the bodybuilder's concern for excellence of construction and correctness of appearance was not confined to private carriages but extended to his more utilitarian productions as well, a fact that can be clearly seen in plates 8 and 9 and I and IV. The overall standards maintained were high but, by common consent, it was agreed that the finest work of all came out of the great London houses. Much of this was the result of the excellent technical training in the trade there, which was to continue through the motor age.

2 *Carriages to cars* 1881 *to* 1914

The fact that the coachbuilding industry was able to maintain such generally high standards was due in large part to the attitudes of those involved; most cared greatly about what they did and how they did it, and this concern was evident in the quality of whatever they made, from dogcart [IV] to dress chariot. After a century or two of continuous development, many of their products had a rightness of shape approaching the limits of contemporary achievement, and a style of decorative treatment, interior furnishing and general finish that was masterly in its bringing together of the best taste, the best materials and some of the best craftsmen of the day. Not all of the work was good, of course; the cheap and shoddy, the flashy and the merely meretricious were not uncommon, for the trade had its small share of rogues who cared much for profit, little for quality, and who were less than careful in their choice of materials and supervision of work done. This placed them on a level with their customers, who cared much more for cheapness than for excellence and who, like them, had no interest at all in the finer points of design, workmanship and finish, or in durability. They deserved one another.

Quality was not simply a question of first cost. Then, as later, excellent vehicles were available at various prices. Good and intelligent design and planning had much to do with this state of affairs; those responsible had intimate knowledge of the properties of their materials, of the capabilities of the men handling them and of the available methods of construction, and were well able to make effective and economical use of these. Techniques were not complicated, on the whole; simplicity was an outstanding feature throughout the eighteenth and nineteenth centuries, with little mechanisation and little investment, therefore, in plant. Hands provided the major contribution, skilfully marking-out, sawing, shaping, bending, steaming, planing, polishing, painting, varnishing, and so on, with help from simple, traditional tools only. Because so little metal was used, the average coachbuilding establishment was a relatively quiet and peaceful workplace. At a later date, however, matters were to change when sheet-metal replaced wood as a cladding material (although its forming was still carried out largely with hands and hammers).

At the top end of the trade, in London, work of the highest quality was produced, equalling the best of the leading cabinet-makers of the times. A rich variety of woods for constructional and decorative purposes was handled with the sensitivity and confidence of outstandingly able craftsmen. Similar skills were exercised by upholsterers and trimmers, whose mastery of their materials, often exotic and extremely costly in themselves, was as complete as that of craftsmen working in precious metals on a smaller scale. The quality of their work was recognised widely and appreciated. At a level where the best was taken for granted, as a matter of course, there was a good deal of discussion between coachmakers and their clients at times amounting to collaboration. The example of Count d'Orsay and Barker & Co has been mentioned; it was but one of scores of similar associations, in the course of which ideas from both sides were pooled, important and interesting advances in design often following. The user, after all, could draw upon practical road experience as well as his store of natural or acquired taste, and the maker could refer to his accumulated knowledge of constructional methods on the one hand, and to his design staff for aesthetic matters, on the other. Practically without exception, however, the designer himself remained unknown; when he was named, it was almost certainly because he was an outsider, a prominent architect, for example, and not a member of the craft. But external assistance of that kind was only sought when coaches of the greatest national, civic or social importance were concerned.

Outstanding among vehicles of this type is the State Coach built for George III, and used for the first time in 1762, when he drove in it to Westminster to open Parliament. It was designed by William Chambers (later knighted), a young architect who had been the King's art master and had several buildings at Kew to his credit, including the pagoda that still stands in the Gardens. The cost of the Coach was considerable; the total a fraction under £7,652, according to A. E. Watson, the writer of the 'Carriages' chapter in the Badminton Library volume, *Driving*. As he said, 'It was in every way as elaborate as it could be made, a circumstance which will be understood when it is said that of the total cost . . . the largest sum, £2,500, went to the carver.' The latter, one Wilton, received more for his contributions than Butler, the builder, who had to be content with £1,763 15s 6d. Pujola, the gilder, was rewarded with £933 14s for his considerable efforts, and the Italian artist, Cipriani, received £315 for his painted panels. The laceman was paid £737 10s 7d, the chaser (of silver fittings, presumably) £666 4s 6d, the mercer £202 5s 10½d (how one would like to see an itemised invoice), the milliner £31 3s 4d and the woollen draper a mere £4 2s 6d. That was not the end of the expense, for repairs and refurbishing during its first century of service were to cost many thousands of pounds. Apart from its size (it is 24 ft long, 8 ft 3 in wide and 12 ft high, and weighs 4 tons) the State Coach is highly impressive, covered and surrounded as it is with applied ornament more notable, perhaps, for richness and exuberance than for aesthetic quality. On the other hand the Coach is extremely showy—a large part of its function after all—and while the writer might not choose to describe it as magnificent many others would. Such a vehicle was rare, of course, which made its visual impact all the more exciting on the few occasions when it was seen in public. Over and above its ceremonial significance it had great value as proof of the creative, decorative and constructional skills of those responsible (English in the main) and of the ability of English coachbuilders to produce a type of carriage normally thought to be a speciality of other nations. To some degree the Docker Daimlers that aroused widespread interest after the Second World War served a similar function. They were certainly in the same tradition of mobile opulence.

The growth and prosperity of the coachbuilding trade owed much to the occasional commissioning of vehicles of such a kind, and to an always well-publicised interest on the part of those prominent in public life; those who made up the relatively small number that could afford the luxury of carriage ownership and use in earlier times. Royal patronage has always been of especial importance in England, beginning as long ago as 1556, when Walter Rippon made a coach to the order of Queen Mary. He built a State Coach for Queen Elizabeth I in 1563, and in 1584 produced a special Chariot Throne for her use. But she preferred to travel on horseback whenever possible, riding in her coaches only when the occasion demanded. She sometimes used a Dutch-built coach that could be opened or closed; open, of course, it allowed its regal occupant to be clearly seen by her subjects, a practical and extremely important point still taken into account by the designers of State vehicles for ceremonial purposes. Today a coachbuilder also has to tackle the additional problem of providing as much inconspicuous protection from attack as possible—a matter probably not taken into account by Hooper & Co. when they built the Coronation Landau [10] for King Edward VII early this century.

10 *This Landau was built by Hooper & Co. for the coronation of King Edward VII. Although each major component is substantially constructed, the fact is well-disguised by excellence of proportion. The design is simple, subtle and strong.*

To be seen was a most important aspect of carriage ownership. However restrained in shape and colour the carriage itself might be, its extremely high overall finish, the presence of liveried coachmen and other servants in prominent positions outside and the sight and sound of splendidly harnessed horses could not pass unnoticed. They were not intended to do so. Such sights have no modern equivalent; only on rare occasions do royal horse-drawn vehicles appear in public, and the over-elaborate Lord Mayor of London's Coach normally makes but one appearance a year. From the point of view of public spectacle this is a matter for regret for, at best, it can be a stimulating experience. The last regular spectacle of the kind that could compare was the Atora Beef Suet wagon that travelled in various parts of Britain before the Second World War. It was a large covered vehicle drawn by a team of white oxen that thrilled innumerable thousands of people as it passed slowly and deliberately through towns and cities, advertising its owner's product in a most attractive and memorable way. It was marvellous street theatre. Much of the appeal of cart, coach or carriage was due to the presence of the horse, appreciation of which involved most human senses, including smell, and was often a moving and rewarding experience. Horse and vehicle were complementary, the one useless without the other for all practical purposes, and careful matching of visual characteristics was of particular importance for public or private passenger-carrying. For goods transport it was less significant, but in practice much interest was taken in the breeding of animals that not only worked well but looked well. The sturdy Suffolk Punch, the proud Percheron and the massive Clydesdale both looked well and pulled well and fortunately they are still to be seen at work in many different parts of Britain.

The better class of privately-owned carriage impressed by its excellence overall; the Victoria is a fine example. Appearance was admirably suited to function; suitability for its intended purpose was as nearly perfect as was possible at any given period, and life-expectation was a long one. Except in extreme cases, carriages stood apart from prevailing fashion as far as basic form was concerned; within the limits of contemporary knowledge, utility and comfort, from the user's point of view, reached high levels, the range of available amenities being surprisingly wide, and fifteen to twenty years of use, if not more, might be expected from a well-made vehicle regularly and conscientiously maintained. But ownership was not essential, for hiring, on short or long-term contracts, was provided by many of the coachbuilders from early times.

The care and maintenance of coaches was quite a complicated matter, let alone the tending of the horses, stables, harness-rooms, and so on. Rusting, of course, was not a serious problem, but there were many others. Although coach-paint looked magnificent when clean and was unexcelled for sheer richness and depth of shine and colour, it was a delicate material for everyday use and had to be treated with extreme care and understanding; in particular, it was badly affected by water or mud and no self-respecting coachman left his vehicle for long undried; wiping-down was an automatic operation after rain, otherwise the precious paintwork was spotted indelibly and the carriage-keeper would soon be seeking another position. Neglect of a highly-finished vehicle was unforgivable [11]. In the coach-house dryness was of the greatest importance; so, too, was fresh air, the latter even more vital than warmth in the well-conducted establishment. The conscientious servant saw to it that linings and cushions were carefully brushed, and if a vehicle were to be out of

I *The mail coach was the long-distance bus of its day; a strong, capacious vehicle that served all the year round as postal carrier. Its crew and several of its passengers were fully exposed to the weather at all times.*

II *This fine-looking vehicle, according to* The Carriage Builders' and Harness Makers' Art Journal *of 1860, was 'an elegant summer carriage for a gentleman's family . . . well adapted from its lightness for a pair of high bred horses, and called after the Honourable the Chancellor of the Exchequer, "The Gladstone"'.*

J & C. Cooper
Copyright

III *The Brougham was a particularly attractive, useful and comfortable vehicle that was popular for private and public service over a lengthy period. The observant will note the dual suspension of this elegant example, by half-elliptic and C-springs, and the unrealistic dimensions of the 'perch' joining its axles* (opposite, top).

IV *The dogcart was a working vehicle, a gun and dog carrier for shooting parties introduced early in the nineteenth century. The horizontal slats admit air to the dogs housed inside the body and steps are provided for the agile to clamber aboard. High ground clearance was an important feature. Some horse-drawn* char-à-bancs *were very similar in appearance.*

V *The look of a late nineteenth-century Benz may have lacked grace but it was not without visual interest; its many-spoked and narrow-tyred wheels (unequal in diameter, like those of a carriage or coach) 'taking the eye' most agreeably. Although a 'perch' still links the axles, carriage-fashion, this is obviously a new and different kind of road vehicle.*

VI *It is a fortunate thing that motorists, over the years, have been able to laugh at their cars and themselves, as those familiar with the pages of* Punch *magazine will know. Early this century the manufacturers of Perrier tonic water published an amusing, well-illustrated book of lithographs pointing gentle fun at man and motor.*

VII *The practice of showing designs to customers in visual form that had been traditional in the coachbuilding trade continued well into the motor age. This wash drawing (circa 1904) is a suggested scheme, by Barker & Co., for a side-entrance body, the Chandos, on Roi-des-Belges lines.*

11 *The Dress Coach (the example shown was built for the Duke of Buccleuch in 1830) was a mobile showpiece for its owner and its builders; a memorable reminder of the taste and fortune of the former and of the skills wielded by the latter. It is especially handsome, finely proportioned, beautifully finished, and enriched in appearance by the elaborate hammercloth draped across the coachman's seat.*

service for any length of time the latter 'should be taken away, placed in holland wrappings together with a handful of Russian leather shavings; while a few more should be placed on the carriage itself to preserve the lining from the ravages of moth.'* He took particular care, also, of the hood and any other leatherwork, which required regular, skilful and sympathetic attention, and a great deal of physical exertion for best results. Maintaining the leather and metal parts of the harness was, perhaps, an even more arduous task, but for those who revelled in the sight of shining 'tackle' the effort was richly rewarding in the end. It was considered bad practice to have harness-room and stable in direct communication 'as the ammonia from the latter dulls and tarnishes all metalwork which may come under influence of its fumes.' The tending of vehicles and horses was a demanding and responsible business, and a highly specialised one, moreover. In the case of the latter, the temperament of the animal itself had to be taken into account. In theory the self-propelled road vehicle avoided this problem but early users were soon to find that mechanical horses could be as difficult, as moody and as unpredictable, at times, as live ones. When mechanical troubles arose, as they did from time to time, nervous occupants of motor-driven vehicles took comfort if they bore a close resemblance to familiar horse-drawn types. Coachbuilders in the early days of self-propelled vehicles took appropriate action [12 and 13].

Passing mention has been made of the steam-driven carriages which had a brief spell of use during the 1830s. There were other abortive attempts to place practical self-propelled vehicles on the road at different times and in different places during last century, but it is generally agreed that the motor industry of the world had its

* *Driving*, Badminton Library, 1889.

12 *The Hansom cab was the forerunner of the mini-cab of the 1960s—a compact conveyance of great manoeuvrability, holding two. It had far more style than the motor-cabs which replaced it.*

13 *Those hostile at first to the idea of a mechanically-propelled vehicle were reassured to some extent if it had a familiar look, a fact clearly recognised by Rothschild et Fils, the Paris coachbuilders responsible for this motorised Hansom on a Renault chassis. The absence of a steering column and wheel remains a mystery.*

beginnings in Germany where, in 1885, Carl Benz and Gottlieb Daimler, working quite independently and without knowledge of one another's aims or activities, carried out their initial experiments. Although this initial lead was to be taken over, for some years, by other nations Germany's worldwide standing in automobile design and construction was to be maintained without difficulty. Nowadays it is at a higher level than ever before.

It was in France that the greatest advances in design were made, during the 1890s and well into the new century. As it happened there was an unusually able, energetic and far-sighted group of engineers and men of business active there at that time, the country had a first-class road system, for which Napoleon Bonaparte had been largely responsible, and its administration took a tolerant view of new forms of road locomotion. Certainly, most of the ideas and the forward impetus originated in France during those early years of automobile development [14 and 15]. Here the general layout of the mechanical side followed the Panhard system (widely adopted even before the end of the century) and in its appearance forward-looking features are to be seen. Progress, inevitably, was relatively slow at first; the mechanical problems to be overcome were formidable. There was also the strength of potential opposition from long-established interests in horse-drawn transport. This was an unknown factor as yet, but with close on half a million horse-drawn vehicles on the roads of Great Britain by the last quarter of the nineteenth century, it was one to be taken very seriously indeed.

There was the matter of horse-breeding and dealing. The animals had to be housed, fed and looked-after, which provided much employment. They had to be supplied with feed of various kinds and provided with harness and other necessary accessories. To increase their working effectiveness metal shoes had to be fitted to their hooves, which made work for farriers, everywhere. On decease their carcases had to be speedily removed and properly disposed of. The manufacture of the vehicles that they pulled was carried out in most parts of the United Kingdom and gave

14 *This Leon-Bollée may have been an earlier version of the car with pointed prow* [14]. *Its all-metal body is held together with rivets and its cooling system is unusually neat and unobtrusive—the radiator, for example, having an almost modern appearance. Note the sprag, angled downwards and towards the rear below the frame amidships, to stop the car from running backwards, and the vertical steering column.*

15 *French automobile designers were the first to investigate the practical advantages of wind-cheating shapes; before the end of last century, in fact. If the evidence of this photograph of* circa 1900 *is to be believed, the spoiler or air-dam, presently popular, is no new thing. In appearance this Leon-Bollée is a strange mixture of ancient and modern. Its side-slung radiator is only one of many intriguing technical features. Marine influences may be detected.*

employment, directly or indirectly, to a great many people. Some idea of the scale of the operation in Britain may be gained from study of early Post Office Directories, which list trades alphabetically and which reach far back into last century.

In Edinburgh, the Scottish capital, in 1881, there were two axle-tree makers, one bridle, bit and spur maker, nineteen coachbuilders, fifty-three coach hirers, two coach-lace manufacturers, twenty-four hay dealers, nineteen horse dealers, seventeen manure merchants, twenty-two postmasters, job horse hirers and livery stable keepers, thirty-two saddle and harness makers, five spring makers (for coaches), two whip and thong makers and numerous blacksmiths. There was also one bicycle maker. Almost sixty years later, in 1939, eleven coachbuilders remained, and by then only one might have had occasional dealings with horse-drawn vehicles, the others all specialising in motor-vehicle work. But there were still seven farriers, three hay dealers, one horse-collar manufacturer, one horse dealer and one horse hirer; eight manure merchants plied their noxious trade and the fact that as many as ten saddlers and harness makers were listed may be explained by the continuing popularity of riding as a form of relaxation and recreation. In Glasgow, a city several times larger than the capital, there were only twelve horse dealers in 1881, and eighteen coachmakers. Besides this there were five horseshoe makers, nine fire engine makers, or their local agents, eight whip and thong makers—and two traction engines. In the same city, in 1914, there were twenty-six coachmakers (the newcomers almost certainly concentrating on motor-vehicle bodybuilding), twenty-eight carriage hirers and twenty-four cartwrights. There were also fifteen motor-cab hirers, forty-three motor-car agents and twenty-seven motor garages.

16 *Only the mechanism was under cover in this 1895 three-wheeled car, possibly designed and built by James D. Roots. Its cycle origins are apparent—in fact it is a powered tricycle, with solid rubber tyres, tiller steering and wire-spoked wheels. It has a flush-panelled body and its rear-mounted engine is well-ventilated inside the perforated metal housing.*

It was this weight of vested interest that any developing, alternative form of transport had to tackle, along with the might of the railways—if they chose to take notice—and the powers of an unhelpful Parliament. The railways were not directly threatened for the foreseeable future, as it happened, and, indeed, were quick to realise the potential practical value of the new type of vehicle as an ancillary in their own business. First of all, however, the road motor had to prove itself.

This took ten years, perhaps. By the mid-1890s it had clearly demonstrated its potential, shown that it could cover considerable distances at steady speeds, with a remarkable level of reliability, in the circumstances. For all that, the 'caprices of the petrol motor' were imperfectly understood, steering and braking systems were erratic in their behaviour and road surfaces were variable, to say the least. For the time being, creature comforts were of little or no importance, and in any case, the physical discomforts of road travel, by whatever means, were widely and readily endured by the majority. It was only the very well-off who could afford to indulge in the most luxurious forms of travel, moving slowly across counties or countries in isolation, cocooned in elaborately equipped and costly vehicles. The open-air aspects of early motoring, seen so clearly in 16 and 17, for example, were not the prime concern of those involved for they were too preoccupied with mechanical problems; too concerned to make their primitive mechanisms work, and work properly at that. The frequent failures, the frustrations, the anguish all were made tolerable, no doubt, by the excitements of success and new discoveries. Nothing stood still for long at that stage of development.

It was by chance that Benz and Daimler arrived at much the same stage in their experimental work by 1885, but there was no chance element in the fact that each

17 *A formal town carriage without horses, the 1886 four-wheeled Daimler, with central engine and directional control by levers, not by reins, was a singularly neat adaptation of an existing vehicle.*

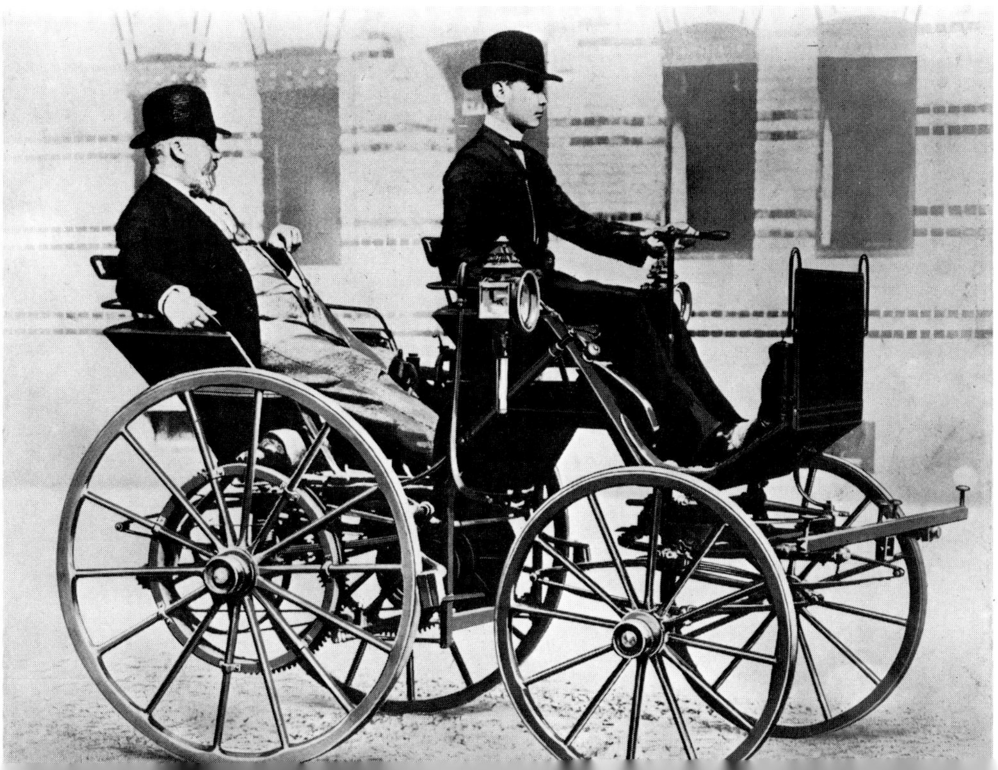

first tried out his particular kind of engine in a road vehicle. For many years Benz had been thinking in terms of a self-contained, lightweight unit to take the place of the horse, and Daimler had in mind the propulsion of road vehicles as one of the many potential applications of his invention. Because each was starting from scratch, without knowledge of the other's activities and with a multitude of problems to resolve (for which there were singularly few precedents) it is hardly surprising that they reached widely differing solutions. The difficulties of designing and manufacturing their new kinds of engine might have seemed sufficiently daunting in themselves although they were much helped by previous experience with the gas engine. What complicated matters was the fact that they had to discover a method of transmitting engine power to driving wheels. There were, as yet no suitable vehicles and it was Benz who was bold enough to create his self-propelled vehicle as an entity. His first machine was, in fact, a tricycle, tubular-framed and chain-driven by a horizontal, rear-mounted single-cylinder engine.

As Daimler was primarily interested in the production of a multi-purpose power-pack he made no attempt to devise a suitable new vehicle but began by fitting his air-cooled, single-cylinder, horizontal engine into a wood-framed bicycle. Perhaps understandably this was not a success, and for a time he and his associate, Maybach, concentrated on engine development. By the spring of 1886, however, they were back on the road, this time with what was almost certainly the first horseless carriage [17]. They adapted a four-wheeler by removing its shafts, installing a makeshift steering gear, and placing the engine in a central position. This, too, was patently unsuitable for further serious development and the associates then followed the example of Benz, quite unwittingly, and produced a vehicle with four wheels and a tubular frame, the latter serving also to cool the water circulating through it from the engine, which was located at the rear. The drive to the road wheels was taken through a gear-train that provided four speeds.

Daimler's main preoccupation was still with engine development, however, and by the early 'nineties he was not only making and marketing his engines on a reasonably large scale but was licensing manufacturing rights to others. Daimler-engined makes were numerous, particularly in France, where the manufacturing rights were held by Panhard-Levassor. This company is generally credited with establishing the arrangement of the main mechanical components of the motor vehicle—engine at front, gearbox and final drive behind it—which has remained the most numerously manufactured.

The Panhard layout [18] immediately influenced the appearance of those makes that followed it and gradually a distinctive type evolved, with front-mounted, enclosed engine behind which driver and passenger, or passengers, sat, far above the ground. It is not altogether true to say that no attention was paid to appearance; as far as time permitted some little attention was given to it, but until mechanisms had developed to a state of reasonable efficiency and reliability it is hardly surprising that the average vehicle had a makeshift air; the result of cross-breeding established carriage types and an altogether new means of propulsion was bound to have an uneasy look [18 and 19]. The Daimler of 1886 seemed to be all wheels, spokes and angularities, without grace or unity. The first Benz, on the other hand, was more

18 *This 6 hp Panhard-Levassor exemplifies its maker's system, with vertical engine at the front, controls and gearbox near the middle and final drive towards the rear. The presence of a screen between front and rear seats is unusual in a late nineteenth-century automobile but the rear-entrance* tonneau *body is not. Note the brake shoe behind the back wheel (a feature of certain horse-drawn vehicles), the cane umbrella case and the awkward appearance of this tall little machine (above, left).*

19 *The Rothschild body sits awkwardly on the high-set frame of this early Leon-Bollée of a semi-forward-control type. The metal panel below the radiator and impressive trio of lamps is not an aid to aerodynamic efficiency but a deflector of dust and other unwelcome objects from the mechanism behind it (above, right).*

20 *The lithe lines and unified appearance of the Lanchester twin set it apart from all its contemporaries. Dr Frederick Lanchester, its designer, had the prophetic, or anticipatory sense possessed by few; a fact insufficiently recognised during his long lifetime. This 1904 12 hp had a wet London to Brighton run in 1953.*

21 *Fine finish was to be expected from a coachbuilder of Rothschild's standing. Here he excelled with this lofty 1898 6 hp Mors, at that time a leading French make. It was heavy, obviously, and its frontal area would put it to great disadvantage against head winds. The use of domed mudguards anticipated a fashion that did not become widespread for at least fifteen years.*

unified; a logical simplicity of conception and construction properly suited to contemporary knowledge, materials and manufacturing methods. Perhaps the finest expression of this was the somewhat later twin-cylinder Lanchester of the early 1900s, which was the first motor-car thought through completely, with the mechanical side and coachwork designed together, rather than as separate entities. It was the modern car in embryo, and the first significant British contribution to automobile engineering [20]. It was not the last.

Although open cars predominated during the 'nineties, largely because engine power was insufficient to pull anything weightier, a sizeable number of closed bodies appeared towards the end of the decade and later. There was a surprising variety of types, a fact not too widely-known as it happens. Pictures in the Institute Library show this variety most clearly. Most of the early examples held there were designed and built by the great Paris firm of Rothschild et Cie, and while CGV, Mors, Panhard and other French makes predominate, several Mercedes are included in the collection [21 and 22]. Until the early years of the twentieth century, France held the leading position as far as the design and construction of automobile bodywork was concerned; her coachbuilders were ahead of all others for elegance and inventiveness, just as French society dominated high fashion. What amounted at times to flamboyance was a feature frowned upon by the more conservative of other nations, yet close attention was paid to new French designs at the international exhibitions, some of which were now taking place in North America. Quite unashamedly, as we have already learned, a party of British coachbuilders went over to the Salon early in the century to find out what the French were up to, adopted and adapted what suited them and quickly took their industry into direct, and successful, competition in the home and overseas market.

Although French makes still retained their overall lead, at that time their position was no longer as sure as it had been. The English Daimler had begun spasmodic production in Coventry in 1897, the first Lanchester car appeared before the end of the century, and other makes were turning tentative wheels, so to speak. In Germany, the Benz (which had become four-wheeled in 1891) remained popular during the 'nineties and found a wide market elsewhere, its reasonable first cost and its reliability making up for a certain absence of life. However, the complacency of its inventor, who moved with the times even more slowly than his little motor carriages, almost lost his company its share of the market, and although Benz survived as a make until its amalgamation with Mercedes, in 1926, it never quite regained its former importance in the industry [v]. As yet there was no strong challenge from Italy where, in 1899, the foundations of the great FIAT concern were laid with the

22 *Neatness and an air of lightness are features of this Rothschild-bodied Mercedes, built around 1902. Points to note are the lack of a windscreen, the well-raked steering column and the double cushion on the driver's seat. Entrance to the rear is gained through the door to the driver's left, an ingenious solution to a difficult problem.*

23 *This attractive 1899 $3\frac{1}{2}$hp FIAT was the forerunner of many millions of cars of the same make (later spelled Fiat), a motorised* Vis-à-Vis *in which driver and passengers sat facing one another.*

appearance of a little 2-cylinder car [23].* The Belgian motor industry also became quite important, the Minerva car and the coachbuilding firm of Vanden Plas, for example, being known internationally and highly respected. In Great Britain expansion began with the beginning of the century and while there is little doubt that the outstanding make, mechanically speaking, was the Lanchester, with Napier, perhaps, a little way behind, the most successful, socially speaking, was the Daimler. This was due largely to its adoption by the royal family.

What had started centuries earlier, when Rippon built the first royal coach for Queen Mary, had continued. The Prince Regent had been a genuinely keen and interested user of horse-drawn vehicles of style and elegance and Queen Victoria had provided the trade with much work during her lengthy reign. Indeed Barker & Co built twenty-four coaches and carriages of different types for her personal use, as well as many others for King Edward VII and King George V. By becoming one of its earliest customers Edward established the English-built Daimler as the 'royal car'. For a time, however, he was an appreciative Mercedes owner, using the big and impressive German car on informal occasions. Elegance of shape was not a characteristic of any of his cars, in fact, but he did take a close interest in their appointments and in their finish; the 'dress' of his road vehicles being as carefully chosen as the clothes and uniforms he wore himself.

The Hooper-bodied Daimler that he acquired in 1904 is an example. In appearance it was rather like a Roi-des-Belges, but the fixed top, extending forward to a curved-top scuttle characteristic of its make, and the enclosed rear compartment were not at all typical of the type. The interior was much more attractive, fitted with

*Alfa-Romeo (originally ALFA) and Lancia were names to be taken seriously much later, their respective heydays coming between the wars and afterwards.

24 *The chassis of this early Panhard is unusually long for its time, yet there is no visible form of additional reinforcement. Period features include a sprag, chain drive, a gilled-tube radiator, very light and simply shaped mudguards and effective decorative use of lining. It is possible that entry to the cosy-looking compartment at the rear, with its luxurious deep-buttoned leather upholstery, was facilitated by tipping forward the back of the front seat—or it may be that the big hand-wheel swivelled the midships bucket-seats out of the way.*

electric light and upholstered in dark-blue buffalo hide, buttoned on the seat cushions, on their backs and round the quarters of the body inside. The wide rear seat could accommodate three passengers and there was space for two more on specially designed revolving chairs opposite. The King, it was said, had supervised their construction. It is not difficult to visualise that interior; rather dark, full of cigar smoke and small talk and very masculine. The external colour scheme sounds unexciting—royal lake, lined with red—but thinking of the depth of colour of the coach-paint and the enlivening effect of much highly polished brass, one realises that the total effect must have been magnificent. Such quality is much less apparent in some present-day cars of state.

Magnificence of appearance was not reserved for royal vehicles. It was easier to achieve in the larger kind of car, where scale enhanced the effect, the area and complexity of forms giving great scope for the exploitation of colour and finish; there were so many surfaces, so many different materials, textures and colours to excite admiration. The Panhard supplied to Baron de Zuylen [24] is a case in point. It was less a feature of the small car, however. In exceptional cases, such a car could be rich-looking and appealing, but always lacked the grandeur and presence of the larger machine. The smallest, the Benz of the 'nineties, the little Renaults or De Dions, for example, the Scottish-built Argyll *voiturette*, the first Rover (the backbone-frame Eight, of 1904), the American Curved Dash Oldsmobile or single-cylinder Cadillac, were simple, or stark, in look. They carried no spare flesh. This was partly a matter of cost and partly a matter of keeping weight down as much as possible, for there was little power to spare for moving superfluous weight.

Although some pioneers followed existing cycle-building practice and used tubular frames, wood, with metal flitch-plates for reinforcement was more commonly used for chassis construction. Neither kind had much beam strength, and so frame lengths and wheelbases remained short until the appearance of pressed-steel frames just after the beginning of this century. In the side elevation the wheels of short wheelbase vehicles were too close together, their bonnets too short, their bodies too high for their length. Nevertheless, many were attractive. Some, to our eyes, look funny—not that that is necessarily a bad thing. Old motor-cars can be taken too seriously, after all [VI].

While the new motor engineers were concentrating their creative energies on much-needed improvements to engines and transmission systems, to chassis design, to braking and suspension, the coachbuilders kept their distance and reserved judgement, which was probably the correct course to follow in the circumstances. By 1900, however, it was quite obvious that the motor vehicle was not going to fade from the scene, and the more progressive members of the trade who had already been involved, looked forward to an increase in business. They were not to be disappointed. From a practical point of view the bodybuilders could deal with any reasonable expansion in demand. Although the motor-car had been under intensive development for some fifteen years its numbers were still relatively insignificant, especially in Great Britain.

At first no drastic changes in working methods were considered necessary but, as time proved, the traditional attitudes of a conservatively-minded craft had to change. Apart from anything else, the kind of vehicle with which they were to be involved on an ever-increasing scale, was very much faster than anything they had previously dealt with and was to set them a great variety of new problems. Bodywork for motor vehicles had to withstand much greater stress and strain; on the whole the open kind stood up remarkably well. Closed coachwork, on the other hand, was more difficult to deal with, and unless extremely high standards of design and workmanship were rigorously maintained trouble soon materialised in use. Squeaks and ratttles became an inevitable accompaniment to the hearty roar of engine, groan of gears and whish of driving-chains. Progress in some early closed cars was not peaceful. However, any concern for quieter operation had to take second place to the continuing search for greater mechanical efficiency and reliability in those pioneering years. Steam- and electrically-driven vehicles were at hand to prove that noise and motion need not be inseparable companions, but the fickle nature of the former and the sloth (and lack of useful range) of the latter made it likely that the internal-combustion-engined car would eventually prevail. Gradual reductions in mechanical noise levels were made but it was not until 1904, in Britain, that Henry Royce showed what could be achieved when the matter was tackled with sufficient seriousness and determination at all stages of design, experimental work, and final assembly. By contemporary standards his very first essay into motor-car design and manufacture, the celebrated 10 hp 2-cylinder model, ran with notable quietness and smoothness—a fact that can be verified to the present day because two of the earliest production Rolls-Royce cars have survived, in excellent working order, and are on the road from time to time.

Reductions in mechanical noise levels had repercussions for the coachbuilder; less

25 *A Daimler draughtsman at work on a full-size body drawing,* circa 1910. *To this day the practice continues, differing only in details and in the translation from two dimensions to three, which is carried out in a different manner.*

and less often was it possible to conceal creaks or rattling in the all-enveloping din. Small wonder, then, that a concern of the high standing of Barker & Co should be named in 1905 as sole supplier of bodywork for Rolls-Royce cars. In practice, however, other firms of comparable reputation were allowed to build for these chassis, and because of the extremely stringent nature of the conditions imposed by Royce himself, by Claude Johnson and by the Hon. C. S. Rolls, whose collective passion for excellence was based on a combination of idealism and practical commercial sense, acceptance amounted to an accolade. Between the wars, the quite large number of 'approved' coachbuilders in Britain alone was a clear indication of the generally high standards maintained across the country. Firms such as Barker, Hooper, H. J. Mulliner and Thrupp and Maberly, in London, Salmons & Sons of Newport Pagnell, Vincent of Reading, Maythorn & Son Ltd of Biggleswade, Cockshoot & Co. Ltd of Manchester, Rippon Brothers of Huddersfield, and James Robertson of Glasgow, were gradually turning more and more of their creative energies and manufacturing skills to the building of motor-car bodywork of high quality, on the best available chassis. Other members of the industry were enjoying a steady, slowly increasing demand for their products that compensated for the drop in orders for horse-drawn vehicles.

From an early stage in the history of the motor industry three ways of dealing with the supply of bodywork were followed. Either the chassis builder concentrated on the mechanical side and bought in bodies from outside specialists [VII], or he supplied chassis to his customers, or he set up his own body-shop, with direct control of its operations. The latter course was impractical, economically speaking, unless it was possible to sustain a reasonably high output.

In England, Daimler was one of the first makers to catalogue complete cars, the quality of work turned out by the company's body-shop being very good [25 and 26].

As a result, demand remained sufficiently high for production to be maintained at the Coventry plant until the 1930s. Because the Daimler was such an 'upper crust' car, however, and at no time particularly cheap, even in its smallest forms, many customers chose to buy a chassis and send it elsewhere, to be bodied by one or other of the specialists. Most of the tall Daimlers favoured by members of the royal family (English perpendicular awheel) had Hooper bodies, for example. Among other English companies that built their own bodywork—and built it very well—were Austin, Humber, Rover, Sunbeam and Wolseley, the period between 1910 and 1914 being one of great activity for all. Apart from Daimler, the only other producer of really high-grade machines to make complete cars on its own premises was Lanchester.

Almost everything that came from that concern was of interest. Dr Frederick Lanchester was Britain's most original creative force in automobile engineering and it is one of the great tragedies of motoring history that the quality and practical logic of his ideas were not appreciated properly in their day. Even now, it is doubtful if his contributions to the industry have been adequately recognised. By the time his first production model appeared in 1901, motor-cars had already assumed a distinctive

26 *Body assembly at the Daimler factory in Coventry around 1910. The bodies seen here are almost ready for painting and trimming, operations to be carried out in separate departments. Traditional methods and tools were still largely-used at that time, handwork predominating as there was still an ample supply of highly-skilled bodybuilders, trimmers and coach-painters.*

27 *This little* 1902 *Napier, with its* tonneau *sheeted-up as a protection against the weather, shows how well a car of the type could look and still looks after almost eighty years of care. Its gilled-tube radiator does not obtrude and its flared front mudguards give it a rather rakish, sporting air. The single, central headlamp makes an impressive figure-head.*

form, modified and distinguished by individual differences in detail and in overall appearance; the majority, even then, had a bonnet at the front, within which stood an engine, and its shape and size—especially the latter—took on considerable emotional significance for some. Before long the untidy and highly-vulnerable gilled-tube kind of cooling system, which occupied a prominent position either ahead of the bonnet or around it [27], gave way to the neater and more efficient 'honeycomb' type of radiator that had been one of the numerous new features of the 1901 35 hp Mercedes, and was soon to be adopted by most other manufacturers. By contrast the 10 hp Lanchester had no bonnet at all and no sign of a radiator (which is not surprising because its centrally-mounted engine was air-cooled). A year or two later, when public prejudice compelled a change to water-cooling, the horizontally-tubed radiator which had to be fitted was slung below the toe-board of the body and slightly behind the tubular front axle, where it was almost invisible. The driver and front-seat passenger sat behind a leather apron to which the front wings were attached, the whole hinging forwards to allow access, and between their comfortable 'chair' and the rear compartment, or *tonneau*, the twin-cylinder engine was located inside a cover, the top of which served as a luggage deck. As a whole it was a design of great elegance, speaking both practically and visually.

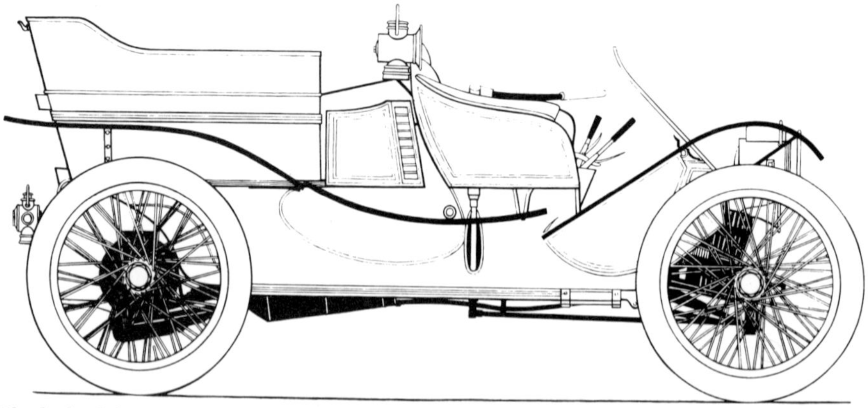

28 *The look of the Lanchester 10 and 12 hp twin-cylinder cars was distinctive, body and chassis being considered as one from the very start of the design process. More than any other automobiles of their time, their appearance reflected the fact that this was a new sort of road vehicle. Because the engine is centrally-mounted between the front seats and the rear compartment (entered, like all good* tonneaux, *from the back), the driver has no bonnet to obstruct his forward view.*

Lanchester realised that comfort was directly related to supple springing, but realised also that this could lead to a loss of control unless an unusually rigid frame was provided. This he secured by starting with frame side-members of aluminium plate, no less than eighteen inches deep, reinforced along their upper edges by an angle girder of steel, to which were fixed wooden attachments for the body. A steel U-section girder along the bottom edges provided further reinforcement also carrying the engine brackets and the trunnions and brackets for the springs and radius-rods. Additional stiffening was provided by a twelve-inch-diameter tubular girder just ahead of the centrally-mounted twin-horizontal-cylinder engine, that served also as the petrol tank and support for the control levers, and by sturdy floorboards, front and rear. The lamp-board across the front of the car was almost an inch thick, and apart from carrying some of the lighting equipment and sheltering the front-seat occupants' feet, it served as bulkhead and scuttle in one. With such a rigid (but lightweight) frame, further strengthened by the fitting of the body, the springs could be allowed great play, yet the patented Lanchester system of twin parallel-links exercised precise control over the movement of the axles. The result was riding comfort vastly superior to that of the majority of cars, not only in its day but for the next thirty years or so. True interchangeability of parts was another feature, applying as much to the coachwork as to mechanical details. Lanchester was his own body designer and as soon as possible became his own bodybuilder as well, making use of jigs for maximum accuracy. This was typical of his logical mechanical thinking [28]. Because he was aware of the importance of ready access to the mechanical parts, he made removal and refitting of the entire body a quick and easy operation; in five minutes or less the 'front chair', luggage deck cum engine cover and *tonneau* could be lifted clear. Other makers used the less elegant method of hinging the entire body structure at its rear end. In the furled position, as it were, it had the look of a sailing vessel waiting for a favourable wind.

Frederick Lanchester was aware of the importance of perfect balance as a positive factor in reducing wear, not only on the mechanism but on the nerves of driver and passengers, and his engines set a high standard in that respect. At a time when the operation of inlet valves was mainly atmospheric he caused them to be opened and closed by mechanical means. He used a pressure-feed lubrication system instead of the drip-feed and splash arrangement favoured by his contemporaries, and his easily-controlled epicyclic gearbox anticipated the self-changing gearbox of the 'thirties and the post-Second World War automatics.

An unconventional approach, however satisfactory in practice, is often commercially unsuccessful, and Lanchester cars, in their original production form, had far too limited an appeal. Their unique side steering-lever control was abandoned eventually. In time, and for similar reasons, the gear-changing, clutch and braking operations carried out so easily, effectively and efficiently by levers arranged on a panel to the driver's left were dropped in favour of a more acceptable conventional scheme, with a steering wheel and minor controls set at its centre. The superb cantilever suspension was retained, however, along with the excellent epicyclic gearbox, and in time the twin was replaced by 4-, then 6-cylinder, in-line engines of even greater refinement. The little French car shown in 29 makes an interesting contrast.

29 *Although this photograph is dated 1899–1900, it is more likely that it was taken in Paris in 1902. The purposeful-looking little car is not overburdened with bodywork, which suggests that its owner was interested in speed. It was the equivalent of the horse-drawn gig, but faster. Typical period details include the weighty gilled-tube radiator, well-louvred bonnet, side gear-change and brake levers, and well-upholstered seat. Final drive by chain was an unusual feature on such a small machine.*

30 *A smart turn-out was appreciated by the keen horseman, who could have found little to fault in this case. The hampers, carried pannier-fashion were, presumably for picnics; the umbrella case beside the rear seat housed portable weather protection for the exposed occupants. In earlier days it would have been a sword case, for protection of another kind.*

Lanchester went his way, other motor engineers went theirs. His immediate influence was limited. The effect however of the 35 hp Mercedes of 1901 was considerable; its mechanical innovations and appearance gave it a special cachet, and its general form and layout were to be adopted, with suitable modifications, by a majority of manufacturers in succeeding years.

With the introduction of this new car Germany regained importance, internationally speaking, and Panhard-Levassor, at that time generally considered to be the most desirable make on the market, had to give best to a model that set the pattern for future developments, not only across Europe but in the United States as well, where, hitherto, automobile design had been in a fluid state, to say the least.

The distinctive 'honeycomb'-type radiator of the Mercedes was a significant advance, not only in terms of increased efficiency for the cooling system but visually as well. At one stroke it tidied up the front-end appearance of the car [30] and provided a positive means of identification. It stood at the forward end of a pressed-steel frame, the intrinsic strength of which allowed greater length (and a longer body) than was possible with the existing forms of construction. These were of two kinds—either the frame was built-up, horse-drawn vehicle fashion, with wooden side-and-cross-members, and was reinforced with metal flitch-plates, or it was assembled from steel tubes, jointed and brazed together in the manner of contemporary cycle frames. The first form was strong but it was heavy and there was an upper limit as far as the dimensions of component parts were concerned. The same applied to the second form, which was strong for its weight but could not be used for anything very large. The tubular 'backbone' type of frame, used by Rover in 1904 and by other manufacturers at other times, even down to the present, is another matter altogether.

Until the turn of the century the practical limitations on chassis length compelled designers to condense the form of a motor-car. Although there was space for entrance to the driving compartment, access to the rear seats, when fitted, had to be made past a moveable passenger seat at the front, or by scrambling in at the rear from a road over which a multitude of horses had passed, more often than not. Obviously this was not an ideal state of affairs. As more and more makers changed to pressed-steel frames, side-entrance bodies became increasingly popular and the rear-entrance *tonneau* and variants passed, quite unlamented, from the scene [31].

Around 1905, the form of what was to become known as the touring car was gradually evolving. Except in the case of the smallest, cheapest machines, it was not yet

31 *A conventional car of its day; a 1903 Wolseley, complete with rear-entrance* tonneau *body, boa-constrictor horn, protective cases over the drive-chains and side brake and gear-change levers. No hood, however, no windscreen, and no instruments visible either, though the English firm of Smiths was early on the scene with special car clocks, then speedometers (along with Elliott, A. T., and Stewart, the American maker). Its dashboard—the vertical board between driver and bonnet—is crowded with trembler-coil box and the drip-feed lubrication system.*

32 *This specially-built Daimler, circa 1906, is untypically long, low and untidy in appearance. Standard productions of the make were more formal and their design better-unified; but the rakish air of this large-engined car is irresistible. Mr Toad, for one, would have found it so.*

either simple in shape or particularly close to the ground, though the example set by Mercedes in that respect was followed by more and more makers. The Daimler shown in 32 is typical neither of its day nor its maker. Height and a compressed profile were characteristic of most motor-cars of the 1890s. Frames were usually simple rectangular structures, set high and commonly level with the tops of the wheels; often there was as much to see below the car as above, with engine, transmission and suspension details, for example, in various stages of mechanical undress. It was not unknown for gear-wheels to be out in the open, while driving-belts and chains whizzed round in an atmosphere that could be unbelievably dusty on the one hand or extremely wet and muddy on the other.

While such a state of affairs prevailed, transmission wear was considerable; gear-wheel teeth wore down or broke off, belts slipped or snapped, chains stretched, came apart, jumped off their sprockets, the latter in a manner familiar enough to cyclists, but dealing with car driving-chains by the roadside was a somewhat different matter. Early motoring was a messy business at times, and it is hardly surprising that so many owners left the care and repair of their vehicles to a gradually emerging band of specialists who doubled-up as drivers, as a rule. Whether or not the motor-car was more difficult to manage than the horse the writer dare not say, his practical experience of equine maintenance being nil. Certainly, it required less attention on a day-to-day basis and, as its champions seldom tired of pointing out, did not eat its head off when at rest. The horse interest could have made much of the undeniable fact that rapid technical improvements caused any model of the car to go out of date with alarming speed, and to lose value with comparable velocity [33].

In Britain, it was not until the turn of the century that the motor-car emerged as a really serious and dangerous threat. Although the total triumph of the car over the horse was to be some time in coming, there could be little doubt that it was inevitable.

In 1897, the chairman of Morgan & Co, a London firm, later closely connected with Adler cars in Britain, declared that his concern was ready to begin car body building despite the universal slowness, prejudice and conservatism of his industry; and much further away, in Aberdeen, local coachbuilders also indicated their interest. In contrast the then-President of the Institute of British Carriage Manufacturers, Alexander Henderson, said that however perfect the motor might become 'most people would still prefer for private use the life-like animated appearance of well-appointed horse-traction to any dead mechanism however smoothly it glided along'. Before long, it is gratifying to note, Mr Henderson's large and important Glasgow firm was busily engaged in bodying dead mechanisms by the score.

From an early stage the name of Mulliner was closely associated with the motor vehicle. By the end of 1898, the well-established Northampton concern, which had begun by operating mail and passenger-carrying coaches and branched out into carriage building, had completed more than 150 car bodies, the majority of which were made to the order of the Coventry-based Daimler Company. The firm (and family) member largely responsible was H. J. Mulliner, whose interest in, and enthusiasm for, the automobile, coupled with great personal abilities and drive, led to a substantial business connection after he joined Mulliner London Ltd, in 1896. In 1900 he was able to buy out his cousin's interest in the motor body side of their concern and established H. J. Mulliner & Co, at 28 Brook Street, Mayfair. Meanwhile the Northampton concern continued trading, and as Arthur Mulliner Ltd, remained in business, and in competition to some degree, with the latter, for many years. Before long the new firm was concentrating on the construction of 'autocar bodies' for De Dion *voiturettes*, Hewetson's Ideal and other Benz and Endurance cars, into which 'aluminium is being largely introduced'. But it was with larger and costlier chassis that

33 *On the road in 1906; the drivers of a Model 'N' Ford and a horse-drawn trap in amicable conversation. This predecessor of the far better-known Model 'T' was a 4-cylinder, 15 hp car.*

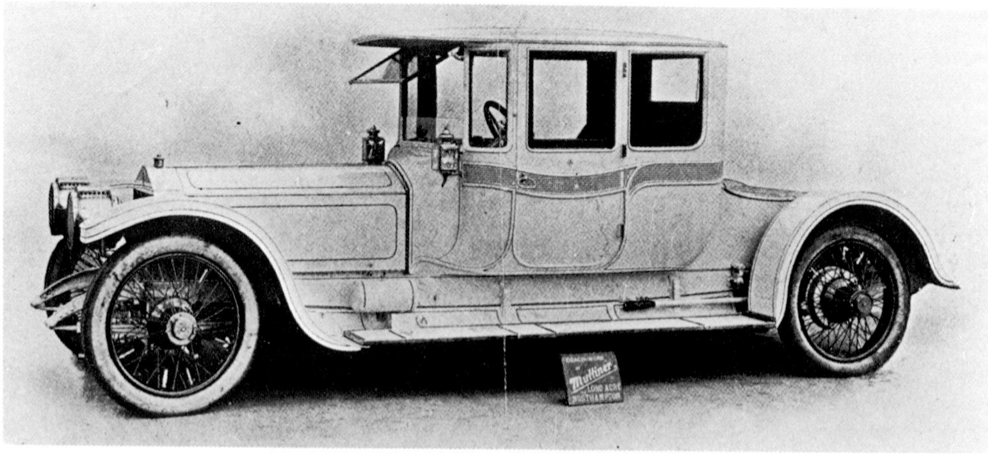

34 *To fit a short coupé body to a 40/50 hp Rolls-Royce chassis, which was long, was one of the most difficult problems for a designer to tackle. Here the person concerned has done remarkably well, anticipating future moves towards greater simplicity of line and form, while making effective use of such traditional decorative details as lining and sham-cane inserts (the latter undulating round the waistline of the body). The pronounced 'peak' of the roof plays an important part in the look of this 1910, Mulliner-bodied car.*

the company established the reputation that kept it firmly in the foremost rank of English coachbuilders. It built for Daimler again, for Napier, for Rolls-Royce and for many others in the luxury class and maintained a level of excellence equalled by few contemporaries [34]. For workmanship and finish a Mulliner body could meet competition from any quarter, and as far as appearance was concerned, internally as well as externally, a large proportion of the bodies turned out from the Chiswick works over forty years or so will stand in motoring history among the finest productions of their period. Even after the Second World War certain Mulliner designs rose far above the unsatisfactory standard that prevailed in the luxury car class until the 1960s.

Because no evidence is available it is impossible to say who was responsible for the consistent quality of H. J. Mulliner work. Certainly the founder had an important influence, and even if he himself was not capable of initiating design schemes on paper, he chose extremely talented body draughtsmen to carry out his ideas. It is reasonable to assume that a strong tradition of excellence was laid down at the very beginning—and certain that very high standards were upheld thereafter. This was so even after Mulliner retired from the trade in 1914, or thereabouts, and as far as is known was not directly involved again, though he was active until the early 1960s. Control of his company was taken over by the Edinburgh firm of John Croall, which had a long connection with the carriage and motor trades.

The individual was still important in those early days, long before any idea of design by committee and computer, or manufacture by robot was in the air. By its very nature the private road vehicle had a more direct appeal than the larger, more impersonal forms of transport, and when its creator had strong individuality, something of this was transmitted to his vehicles—and was liked, or disliked, by those who used them. Of the individuality of the Lanchester, for example, there could be no doubt. In appearance and in the manner of its control, apart altogether from the manner of its going, it was out of step with contemporary thought, and was not commercially acceptable until quite drastic changes were made to its outward appearance. Yet the later Model 'T' Ford, which was as unconventional in many of its mechanical details, and temperamental, too, gained immediate acceptance, not only in America, the country of its origin, but on a worldwide basis within a year or two of its introduction; and the 'coal-scuttle' bonnet of Renault cars does not appear to have affected sales for something like thirty years [35]. In our own day, the idiosyncracies of Citroën design thinking, and a Gallic perversity as far as appearance design is concerned, have had no adverse effect on commercial acceptability and success. In spite of what the design pundits may say, good design does not automatically guarantee good business. But good design in their sense often has more to do with aesthetics than anything else, and when practical advantages predominate, as they most certainly did in the case of the Ford and do, as far as present-day Citroëns are concerned, looks can take a secondary place.

35 *The 'coal-scuttle' bonnet and radiator-behind-engine arrangement retained by the French Renault company for almost thirty years had many imitators at one time. This 1907 example was bodied by H. D. Cleland, of Edinburgh—a concern that took up one of the earliest agencies for the make in Britain in 1902. Its founder is at the wheel.*

This is not to deny the importance of high qualities of appearance which, when properly combined with mechanical design, can be enormously satisfying. This seldom happens to begin with; the 35 hp Mercedes had a clumsy look at first, but within a very short time subtle adjustments to shape and proportions noticeably improved the design. Indeed, certain early Mercedes models and their 1914 Grand Prix cars, for example, possess an excellence of appearance (apart altogether from proven mechanical quality) that gives them classic status. The same cannot be said of Napier cars which, for some years, held a position in English motor-car circles comparable to that of Mercedes abroad. The appearance of these cars was at no time

36 *To provide additional passenger space, the 6-cylinder engine of Napier cars was set forward in the chassis, a fact that was fully exploited by Barker & Co. Built* circa *1910, this body is beautifully finished, but sits uneasily on the chassis.*

an admirable feature. The reason for this was the large and not especially attractive radiator which stood far ahead of the front axle, set there to allow room for the long 6-cylinder engine (an unusual feature to begin with, before 'multi-cylinderism' became fashionable) without encroaching on passenger space behind the scuttle. It dominated excessively and the balance of the entire car was wrong. There was little that any coachbuilder could do to improve matters visually [36].

As to the mechanical side of the firm, Montague Napier was a more than competent engineer, whose involvement with the motor-car soon drew him into a close and, latterly, a not always tranquil, association with S. F. Edge. By 1900, Edge was already an old hand as far as motoring was concerned, with a special interest in, and aptitude for, competition work. He was an accomplished salesman in the grand manner, with little, if anything, to learn from the most aggressive American. Edge never let up in his campaign to promote the Napier car and the 6-cylinder principle, even though his company was by no means the first to adopt it. Apart from frequent use of the correspondence columns of the motoring papers to advance his views he gave great importance to competition participation as a means of gaining attention on the widest scale. He succeeded. In the early days the effectiveness of special exhibitions was limited; there was little to show and not enough interested people (because they were not knowledgeable enough) to go along. Public demonstration, particularly if a competitive element were present, was of much greater value, both directly and indirectly. A victory had great publicity potential that could be speedily exploited in print or by word of mouth. Success—and failure, for that matter—could be valuable for the engineers in their ceaseless search for higher efficiency, better performance, greater reliability and so on. The maxim that racing improves the breed is only valid as long as advantage is taken of the lessons learned. It is of great practical value for speeding-up development.

Road racing in France began in the mid-1890s, and while it may not sound impressive to our ears, the fact that a Panhard averaged 15 mph from Paris to Bordeaux and back, in 1895, was rightly considered to be a tremendous achievement. The distance covered was over 700 miles: the claimed output of the engine of the winning car, no more than four horsepower. By 1900 it was possible to take a Benz to 20 mph under ideal conditions and a Panhard to twice that speed, in each case with more than an element of risk because of low standards of, what is now known as directional stability. In 1899 Jenatzy had covered a flying kilometer at more than 65 mph in *La Jamais Contente*, his strange, cigar-shaped, electrically-driven machine. It was the twin elements of speed and reliability that were all-important, and in the series of competitive events that took place, largely on the Continent, those qualities were being developed fast. So, too, were such vitally important features as suspension and roadholding, steering and brakes. Perhaps the best way to gain an idea of the conditions under which early racing took place is to watch a set of films compiled by the Shell Company during the 1950s, and in particular the one entitled 'The Heroic Years', which covers Grand Prix racing between 1901 and 1914.

In the early years a strictly enforced weight limit compelled designers to cram the largest possible engine into the lightest possible chassis, power output at that time being almost entirely a matter of engine capacity. Engines of 15 to 20 litres or more were a commonplace and speeds were frighteningly high, yet designers contrived to keep pace in the field of steering and roadholding. Close study of the earliest racing sequences shows that stability was surprisingly good, despite the fact that shock-absorbers were not then in use. By 1908, or so, when damped suspension was fairly universal, the ability of racing cars to corner 'flat', at high speeds, was impressive. Abroad, of course, there were no speed restrictions to hinder the designer and, although the use of public roads was controlled to some extent, official attitudes were certainly free and easy in comparison with those that prevailed in Britain. The so-called Red Flag Act had been repealed in 1896; from then on road motors were allowed to move about freely, without the attendant human hindrance fore and aft, carrying a red flag, and the speed limit was raised provisionally to 14 mph. It was then dropped to 12 mph by a spoil-sport body known as the Local Government Board. At a later date it was raised to 20 mph, a not unreasonable figure for driving in crowded areas, early twentieth-century braking systems being what they were. Needless to say, it was not observed and stories of police persecution of the motorist abounded.

It was well for the motor movement at large that so many individuals whose imagination had been excited by the idea of the self-propelled road vehicle were men of social consequence; literate, articulate, forceful yet diplomatic; able to present a reasoned case for repeal of an unrealistic Act and see it through to ultimate success. With the brakes off, so to speak, it was the turn of designers and engineers in Britain to make what they could of the new form of transport. There was nothing comparable to the railway boom of earlier years but the notorious Hudson, whose monetary manipulations had enlivened the development of rail companies, had a counterpart in the equally notorious H. J. Lawson, who made a bold attempt to monopolise motor-vehicle manufacture in this country. Although Frederick Simms had acquired the English rights to the Daimler patents in 1891, he was persuaded by Lawson to part with them for a very handsome profit; as we have seen, production of the English Daimler began, somewhat tentatively, in 1897, in the midland city of Coventry. It was a most suitable place for the purposes of motor manufacture. For one thing it was advantageously situated in a geographical sense, with excellent lines of communication to all parts of the country. For another, there was a ready supply of experienced workmen whose skills had been acquired in cycle building over a lengthy period. Earlier in the century the city had been noted for its ribbon-weaving activities and for watchmaking, but labour problems, foreign competition, tariff restrictions in certain foreign countries and short-sightedness on the part of management had practically killed-off the former, and cheap watches from other countries, in particular from the United States, effectively closed down the latter.

Revival of the city's industrial fortunes began early in the 1860s, when the manufacture of sewing-machines was begun by Josiah Turner, James Starley and an American called Salisbury, whose firm, the Coventry Sewing Machine Company, was founded in 1861. In time it turned its attentions to the manufacture of velocipedes based on the French Michaux pattern, but because its articles of association did not allow for the making of wheeled vehicles they had to be altered; soon afterwards the concern was renamed the Coventry Machinists Co Ltd. In 1870 Starley left, to set up on his own account with a friend and colleague, William Hillman. Out of their association emerged, eventually, the Rover bicycle, the Rover motor-cycle and the Rover car. Although his main interest was the bicycle, in two- or three-wheeled form, Starley did produce at least one self-propelled vehicle before

37 *All was in order at the Longbridge, Birmingham, works of the Austin Motor Company when this photograph was made a year or two before the First World War. Austin bodywork was always well-made and good-looking, Herbert Austin himself playing a dominant part in its design. Note the test chassis on the extreme left.*

38 *The Doctor's coupé was a very popular type, both before and after the First World War. Unusually, this Austin could carry three, a little occasional seat being placed before the front-seat passenger. The interior is trimmed in Bedford cord, a material long popular for closed cars. Note the roof ventilator, or skylight, and the 'railway-carriage' straps for raising or lowering the windows. Luggage could be carried on the railed deck at the rear.*

1890 a rather racy-looking electrically-powered tricycle that was shown in early Rover Company car catalogues with the caption 'The First Car built and tested in Coventry, in 1888, by the late J. K. Starley, of Starley and Sutton, the founders of the Rover firm'. During 1903 *The Autocar* reported that Starley had tried out his interesting vehicle in France, at Deauville, free from speed restrictions. However, it is hard to believe such a story since the rear-motored Rover could scarcely reach 8 mph. Its somewhat spartan-looking body was made by the Coventry coachbuilding firm of Hollick, which later played an important part in supplying local motor manufacturers; and, as Hollick and Pratt Ltd, made coachwork for some of the early Morris cars, eventually being taken over by William Morris in 1922.

The Rover story is typical of the way in which many car firms began. The cycle boom of the mid-1890s had been succeeded by a spell of over-production and by a number of failures, so it was logical for would-be manufacturers to take advantage of the increasingly widespread interest in the motor-car. What Lawson had started, when he established the Daimler Company in an enormous, disused cotton mill in Coventry, had significant consequences for that city, which remained the main centre of motor vehicle manufacture.

The nearby city of Birmingham was also involved in the motor industry from an early stage. Frederick Lanchester began his first experiments there, and his company was based in Sparkbrook until its absorption in the Daimler/BSA complex in 1931. The Wolseley concern was the largest motor firm in the city until it was overtaken in terms of size and output by Austin, after the First World War. The scale of operation even before the First World War is shown in [37]. A typical Austin is shown in [38].

The Rover Company opened a branch factory there, in 1919, to make its new Eight in quantity, and BSA and Singer were other makes with Birmingham plants. London was always more important as a centre for coachbuilding, but was also the home of Napier and Talbot cars, to name the most famous of their day and, between the wars, of the equally celebrated Bentley. Elsewhere in the country there were numerous individual factories. In Manchester it was inevitable that the long-established Crossley concern would turn from gas-engine manufacture to the motor-car, and at a later date to the lorry and the bus. That city's most famous production was the Rolls-Royce car, which was developed there between 1903 and 1904 and went into small-scale production towards the end of the latter year. Its quick commercial success, founded on excellent design and construction, delicacy and precision of control and far-above-average levels of quiet, smooth running, led to a search for larger premises, ending in 1908 with a move to Derby, where car production was maintained until 1939. Since then it has been carried out at Pym's Lane, Crewe.

By 1904 the British motor industry was firmly established, and was expanding in an encouraging manner. Napier racing successes at home and on the Continent had given the make international recognition; the Daimler association with the royal family was of great commercial benefit; Lanchester was making motoring history, without possibly knowing or caring; and a rapidly growing number of other firms—Crossley, Hillman, Humber, Austin, Wolseley, Singer, Star, Sunbeam, Talbot, Vulcan and Thornycroft among them—had promising futures.

So far all the emphasis has been placed on private car development because it was by far the most important as far as Britain was concerned, between 1895 and 1905. During that period public passenger transport was a matter for the railways and for the bus or tramway car powered by electricity, or horse-drawn. The horse-drawn bus, which had been employed in London since the early years of the nineteenth century, gradually gave way to the horse-drawn tram, and then to the electric tram, in its turn the latter faced competition from the motor-bus, which had the advantage of much greater flexibility of movement. Steam-driven buses had been tried out unsuccessfully in London in the late 1890s. Although their use was to be resumed there and elsewhere in the country after 1900, they were too troublesome in service

39 *The first Halley commercial vehicles were steam-driven, but by 1905 their manufacturer was only making petrol lorries and buses at Yoker, near Glasgow. The 1911 tipper seen here was built especially for Liverpool Corporation, local authorities in this country being early to recognise the potential of the self-propelled vehicle.*

40 *The open-topped bus lingered on until the 1920s, in London, where Metropolitan Police control (which still extends to taxi-cabs) was unnecessarily conservative and restrictive. The driver of this 1905 Leyland-Crossley is well-protected, unlike his outside passengers above.*

to offer serious competition to the motor bus, once it had achieved an operationally acceptable standard of reliability, by 1910, or thereabouts. Goods transport over short distances remained largely in the hands of the horse interests until the First World War, though many municipalities made use of the motor vehicle [39]. Again, the achievement of a tolerable standard of mechanical reliability delayed widespread acceptance, and although the number of motor lorries in use began to increase notably towards the end of the first decade of the twentieth century, it was not until the 1920s that it rose to significant levels. Before then there was little difference between goods and passenger chassis; their frames were high-set, flat-topped, carried on solid-tyred, cast-steel, spoked wheels, and propelled by large, low-powered, 4-cylinder petrol engines of straightforward design [VIII and IX].

One notable exception was the Daimler KPL bus chassis that made a fleeting appearance in 1910. It featured two 12 hp Knight sleeve-valve engines that were slung below the main structure, one to a wheel, forming part of a Pieper petrol-electric transmission system, Lanchester's patented worm final-drive, four-wheel brakes and a combined underframe and body that formed an all-steel unit. With its tall, wire-spoked wheels it had a rather frail apperance, but the concept was an imaginative one and it is a pity that litigation on the part of Thomas Tilling, claiming infringement of petrol-electric transmission rights, prevented the formation of a London bus company to operate a fleet of these technically adventurous vehicles. In 1898 Radcliffe Ward had obtained a London Metropolitan Police licence to run a battery-powered electric bus, but the type was patently unsuitable and no more was heard of it. The first petrol bus ran in London in 1899. Even before then, however, in other parts of the country, motor wagonettes had been tried for public service work. In Edinburgh, for example, the first licensed service began in 1898, with Daimler and the Motor Manufacturing Company (MMC) vehicles, but it lasted

41 *The Scottish firm of Argyll was versatile, turning out cars, taximeter-cabs, goods chassis and buses from its impressive, sandstone-fronted factory near Loch Lomond. This 1908 charabanc—to use the English form—carries bodywork of minimal type, and was colloquially known as the 'toast-rack'. It has a strange system of front springing and its substantial engine is placed between driver and passengers.*

no more than three years. Almost certainly a major factor in its lack of success was the unreliability of running. The same problem caused the Birmingham and Midland Motor Omnibus Company to give up its motor buses in 1907, in favour of horse-drawn vehicles, and it was not until 1912, when much more reliable types of motor bus were available, that the concern renewed its trust in them. Some typical pre-1914 types are shown in plates 40, 41 and 42.

In London, where neither the steam nor the petrol bus had been without problems, and where there had been a multitude of different makes and kinds in use to complicate matters, rationalisation was the only solution. The first standard bus, the 'X' type, went into service towards the end of 1909, its design incorporating the best features of the twenty-eight types that had stood up well to the notoriously trying conditions of metropolitan operation. The 'X' type, and its greatly improved successor of 1910, the famous 'B' type, were designed by Frank Searle, then the Chief Motor Engineer of the London General Omnibus Company. Of the 'B' type almost 3,000 were built, the majority of which had a 30 hp 4-cylinder engine, a three-speed gearbox and worm final drive, and a double-deck body, with open top. Solid tyres were fitted and the driver sat behind an imposing bonnet. It was a rather badly proportioned vehicle. The post-war 'K' type was a great improvement, visually speaking; the driver was placed alongside the bonnet, Lanchester-fashion, on the offside, with a splendid view forward—an important consideration in the more crowded conditions that prevailed from 1919 onwards. It was probably the best-looking bus then in service and the high standards of cleaning and maintenance upheld by its owners meant that it was always well turned-out.

42 *Gradually more thought was given to the comfort of the passenger in public service vehicles, the difference between the Halley coach shown here and the Argyll charabanc being noticeable. As further protection roll-up, roll-down side curtains were provided in this 1913 example.*

43 *The Swiss Martini was popular (on a limited scale) in Britain in the early years of our century. Thorn, a London coachbuilder, was responsible for this restrained touring body in 1905. Interesting features are the oval-shaped acetylene headlamps, the combined arm-rest and storage locker dividing the front seat and the neat step below the rear door, which served also to hide the forward chain sprocket and chain of the final-drive system.*

In the period under review the appearance of all road vehicles changed greatly [x]. In 1900, the horseless carriage predominated; with few exceptions it looked what it was, a powered cart or carriage. By 1905 the private car had developed a distinctive form, with bonnet, scuttle and passenger-carrying space behind [43]. At that time the vertical scuttle, or dashboard, was widespread but it gradually gave

44 *The division between the bonnet and body of the 1906 Martini is abrupt. By 1912, when this French Vermorel was built, engine and passenger compartments merged smoothly, so that the look of the car had greater unity. The two-seater was an extremely popular body type, with auxiliary seating in the 'dickey'. This example was made by Mills, of London (above, left).*

45 *The abrupt separation of low bonnet and tall scuttle suggests that this Crossley (from Manchester) dates from 1911 or 1912. The smooth shape and full curves of its body indicate that it is of later origin, possibly 1913 or even 1914. The spare tyre is of the steel-studded pattern commonly thought to reduce the danger of skidding.*

46 *An invasion of foreign cars at the Scottish Motor Show in 1910. The Renault chassis is substantial and an excellent foundation for well-made coachwork. It features its maker's characteristic radiator-behind-engine layout, a deep undertray to keep dust and mud out of the mechanism, and shock-absorbers.*

way to the taper scuttle, which made the transition between bonnet and body less abrupt and emphasised the horizontal aspect in side elevation. From about 1911 onwards, the simplifying of body panel shapes, which were flattened-out, and the raising of body sides gave a smoother, cleaner shape that was more unified than previously [44]. From a practical point of view, these simpler shapes were less difficult to make, easier to paint and finish and maintenance was a good deal less time-consuming [45]. Not that that was a matter of personal concern to the motorist who had a 'man' to drive his car and look after it on a full-time basis; the car-wash of those pre-First World War years was a paid man with a sponge and a chamois-leather, an unlimited supply of water and a few hours to spare.

As the British motor industry grew in size and importance after 1900, the import of foreign cars continued [46 and 47], but it was by no means a one-way traffic. In 1908, for example, 757 cars came into Britain from the United States, by then the largest automobile-producing country in the world; in return, sixty-five British cars were sent across the Atlantic. The best year for British exports to America was 1912, when the total was 187. Although such figures may seem ridiculously small it should be remembered that the majority of British vehicles sold in the States would be of the costliest kind, bought for prestige, more often than not.

As far as Europe was concerned, use of the automobile was largely confined to the better-off, who appreciated to the full its practical advantages for short- or long-distance travel and the status it conferred. That was not so in North America, where the utilitarian value of the automobile was highly regarded almost from the start of the motor age. Operating conditions differed greatly from those prevailing in much of Europe and driving distances in the United States were enormous by our standards. The typical American car that evolved was roomy, softly-sprung and powered by a large-capacity, low-efficiency side-valve engine. Its suitability for home conditions

47 *A leading Scottish coachbuilder was James Robertson, of Glasgow, who had early connections with IBCAM, and whose firm made many fine bodies for the better kind of chassis. A typical example of his work is this 1912 Panhard-Levassor* landaulette.

automatically fitted it for universal use, and a large export trade was the lasting result. The apparent flimsiness of some American automobiles was deceptive; in fact they had great reserves of structural and mechanical stamina. They needed them. Away from urban areas, roads in the well-developed European sense simply did not exist. Conditions were indescribable for much of the year, and the experience gained after a century or two of horse-drawn vehicle use was consulted. The motor-buggy was one solution that quickly developed; the little 'curved-dash' Oldsmobile of 1902–1904 was a typical example. It was very light, its pneumatic-tyred wheels were of quite large diameter and consequently ground clearance was high; and an extremely low bottom (or 'low' gear) enabled it to crawl through all but the deepest layers of mud. The 'high-wheeler' was another answer to the problem, and it was practical enough for its intended purpose to remain in production until 1914, or thereabouts. It, too, was a motorised version of the horse-drawn buggy and looked it; its wheels were of very large diameter, of the rubber-tyred carriage pattern, and its ground-clearance was exceptional. It was sometimes known as a 'swamp-buggy', for good reasons. The original motor-buggies had their natural successor in the Model 'T' Ford and it, in turn, was followed eventually by the Jeep and its derivatives, with their exceptional off-road capabilities. The Ford was not the only American car specially suitable for home driving conditions; a distinctive type quickly evolved that was large, amply-powered, sturdy and untiring to drive. In a country where fuel was so plentiful and so very cheap the thermally inefficient, large-capacity engine, with its high gasoline consumption, made practical sense.

Nevertheless, Henry Ford's Model 'T' was by far the most successful. Its combination of low running costs, lightness, sufficient power, durability and versatility, with high quality materials, and construction, was unique. Because of its two-point suspension, its light weight, its high ground-clearance and its ample low-speed pulling powers it was a go-anywhere vehicle, ideally suited to off-road motoring. The simple, robust mechanism was easy to understand and to repair, and spare parts, which were inexpensive, could be bought almost anywhere. At the price, perfection could not be expected; the Model 'T' was tantalisingly temperamental at times and its performance, in measured terms, cannot be described as sparkling. Its appearance remained aloof from passing fashions: strangely neutral always, but rather more appealing after the change to a rounded-top radiator during the First World War. Above all, the Model 'T' introduced millions to the motor-car (there were some who denied that it was indeed such a thing . . .) and played an important part in opening up America. It was, also, the first masterpiece of production engineering, the first great achievement of automobile mass-production, although it was in no way the first mass-produced car; many years earlier, in 1903 and 1904, Ransom E. Olds had turned out several thousand Oldsmobiles by intelligent use of techniques of quantity-production initially developed in the American gun and watch-making industries.

In an endeavour to meet demand and to avoid import duties, manufacture of this highly individual machine spread inevitably to other countries, England among them. At first, in 1911, production, or rather assembly—for finished parts were shipped across the Atlantic by the boat-load—took place at Trafford Park,

Manchester, in a former tramcar body-factory. By 1914 Ford had almost twenty-nine per cent of the British home market. In that year the number of motor-cars in use in Great Britain was over 132,000. There were 82,000 goods vehicles on the road, 51,167 hackneys (a word that covered public service vehicles as well as taximeter-cabs) and countless motor-cycles, tramcars and horse-drawn vehicles. Most were petrol-driven; competition from steam and electricity had not developed as some had expected. Once the internal-combustion, petrol-powered engine had reached an acceptable level of efficiency, reliability, flexibility and quietness, the battle was won. When the self-starter became a reality, a year or two before the outbreak of war in 1914, the scope of the petrol car was greatly enlarged. No longer did the lady motorist have to possess an unladylike physique for twirling the starting-handle, and female involvement in driving increased rapidly.

Although management of the electric vehicle was a simple matter, and its quietness, smoothness and freedom from fumes or other mechanical manifestations had a strong appeal, a severely limited range made it suitable only for town or city work. Its charm was undeniable; socially it was *chic*; and from an operational point

48 *This 1905 White 15 steam-car has an 8-foot high, rear-entrance* tonneau *body built by Cann Ltd, of London. Outwardly its appearance is that of a petrol-car but what seems to be a radiator is in fact a condenser, and engine controls differ considerably from those of contemporary internal-combustion-engined vehicles. Along with the Stanley and its derivative, the Locomobile, the White was popular in Britain. All were American.*

of view the fact that vehicles were readily available in London on a contract-hire basis (with driver, if required), meant that problems of maintenance and repair were dealt with by experts. The main disadvantage, however, was the electric storage battery, which was large and extremely heavy and demanded frequent recharging. Its weight also adversely affected tyre life. The type, which lingered until the 'thirties in America, is currently being considered for development again. For all the practical advantages of the steam engine, it was too inconsistent in its everyday behaviour for anyone but a skilled, sympathetic and sensitive engineer to manage [48]. Its quietness of operation, its marvellous flexibility and its ability to leap away from a standstill had appeal, but fits of temperament, constant demands for water and paraffin or petrol for the burners, and a shameful tendency to leak oil, water and steam in public made it a less attractive propositon for the majority. Almost anyone could learn to master the basic mysteries of the petrol car. This was certainly not the case with the steam-driven car and the type, which had never been greatly developed in Europe for private purposes, was effectively off the market well before 1914. It took longer to disappear in the United States. In the early 1920s there was an attempted revival, with Stanley prominent for a time, again, and then the technically advanced Doble. Cars of both these makes reached England and were well-received, but they gained no significant commercial success, and although further occasional efforts were made to resuscitate steam for private car, bus and goods-vehicle propulsion well into the 1930s, its day was also over, by then, in North America.

In commercial use, however, steam power was still viable. Reference has been made to the failure of the steam bus. Steam-propelled goods vehicles had a much greater chance of success, largely because their operating conditions could be more readily and easily adapted to suit their working characteristics, and a number of excellent makes remained on the market, both before and after the 1914–18 war. The Lancashire Steam Motor Company was established in 1896 to manufacture steam vehicles, and its 30 cwt van, which had an oil-fired boiler and a 2-cylinder compound engine of 10 to 14 hp, won a Gold Medal at the Manchester Trials of 1897, organised by the Royal Agricultural Society. The high reputation of the company, marketing under the trade name of Leyland, was an asset at a later date, when a change was made to petrol-driven goods vehicles and buses. Well before 1914 these Leylands were among the best of their kind.

A contemporary, and rival, was the Thornycroft Steam Wagon Company, which was also founded in the 1890s and made a comparable name for the quality of its productions. The company changed course—appropriately enough for a concern with such strong marine connections—in 1907, to concentrate on petrol vehicles [49]. Before then, however, it had been responsible for some excellent designs and for at least two historically important innovations. One of these may well have been the first-ever articulated road vehicle, and the other, according to E. L. Cornwell,* may have been the world's first light goods vehicle, conceived as such.

The Sentinel, which had its beginnings in Glasgow, a city closely connected with steam power for marine and railway use (and with James Watt at an earlier stage), was one of the best goods carriers of its kind ever made in this country; it was to remain in existence as a make almost longer than any of its contemporaries. In design it was of the undertype persuation, as it were; its engine was mounted on the chassis and its boiler, like most of the mechanism, was out of view. The Foden was another famous steamer and just as distinctive in appearance as its Scottish rival; it, too, was highly regarded. It was of overtype form, with a long, locomotive-type boiler surmounted by the engine and topped at the forward end by a tall smoke-stack. With mechanical parts whirring round or reciprocating in full public view, as well as a

* E. L. Cornwell, *Commercial Road Vehicles*, B. T. Batsford Ltd, 1960.

49 *This is not a petrol tanker of the specialised type developed after the First War but a makeshift carrier of lamp oil for its owners, the Anglo-American Oil Company (now Esso Petroleum). The chassis is a 1904 Thornycroft, with a special power take-off from its petrol engine to drive the tank pump.*

crew visibly busy at all times, its progress was a matter of interest and excitement to a great many small boys through the 1920s and well into the next decade. By its very nature the undertype was much less dramatic in action, but it also had its admirers. The shape of later Sentinels and Fodens was full and satisfying; business-like in a neat and tidy manner. There was always the chance of a wisp or two of steam or a dropped, smoking cinder to give the game away; with the later, well-developed types, on pneumatic tyres, the source of power was as unobtrusive as that of a trolley bus or a Rolls-Royce.

Such standards of quietness and flexibility, with none of their practical disadvantages, were realised by Henry Royce in his 40/50 hp petrol-engine model, first shown publicly in 1906. It was a marvellous achievement for an engineer who had begun to take a serious interest in the motor-car only three years before. His previous experience had been in the electrical industry, where the small firm of Royce Ltd had established a reputation for the excellence of its products, for their reliability and durability and for the quality of their construction. Spurred to a passionate pursuit of the highest standards by an innate idealism, and convinced that he could make a much better job of car design and manufacture than most of those already involved, Royce created his first model, the 2-cylinder 10 hp, in a

50 *Unlike so many of its noisome contemporaries the 1904 10 hp Rolls-Royce moved as unobtrusively as a well-trained servant. This, the oldest known survivor, may have been the demonstration car at the Paris Salon of December 1904, when the make was first shown publicly. More than seventy-five years later the shape and basic proportions of the best-known radiator in the world remain unchanged.*

matter of months [50].It is said that the prototype Royce car ran on the road on April Fool's Day, 1904. From all accounts it performed convincingly, and before long two more experimental chassis had been completed and joined it for intensive development testing. The story of the meeting in Manchester between Royce, a mature engineer but a newcomer to the motor-car and Charles Rolls, one of the most experienced and knowledgeable motor men alive, is a familiar one. It is significant that Rolls immediately appreciated the special qualities of the little car and wasted no time in securing sole marketing rights for future products of a company with no previous experience of motor manufacture. The Manchester meeting early in May 1904 was arranged by Henry Edmunds, a friend and business associate of each of the principals involved, well-aware of its potential importance for both sides. In taking on, what was for him, a largely unknown quantity, Royce had risked much, not only for himself but for his modest company. All the same, the electrical trade was in a depressed state in 1903; the home economy was still recovering from the adverse effects of the recent Boer War and strong competition from abroad, based on drastic cost-cutting, was making life difficult in more than one British industry. Diversification, especially into an expanding trade, seemed a possible solution to prevailing problems. There was a strong element of speculation in the project, and

Royce did not find it easy to persuade his fellow-directors to sanction its beginning. All were worried about the future and about the commercial survival of their concern, but they listened, they considered and finally agreed to his proposals. Perhaps they complained from time to time about the apparent extravagance of his painstaking experimental work, but they knew their man and their trust was fully justified within an unexpectedly short period.

As far as Charles Rolls was concerned, Royce was also something of an unknown quantity. The idea of a north-country electric-crane and dynamo maker turned car designer must have sounded slightly improbable, even to him, but he trusted Edmunds and he was very anxious to find a British car of real quality to market through his retail business, C. S. Rolls & Co, in London. Although he was mainly concerned with the Belgian Minerva and the French Panhard-Levassor at that time, and dealt in many other makes—mostly foreign ones—his declared aim was to sell British. It would be interesting to know just why it was that Rolls did not handle the best available English makes of the day—Daimler, Lanchester and Napier. One has to assume that he was unwilling to do so, or that he was unable, because of the marketing policies and arrangements of the companies concerned. Perhaps his conspicuous lack of success with Wolseley 'Beetle' racing cars had soured his attitude towards home productions. The fact remains, however, that he was sincere in his wish to find and promote a British car of appropriate quality. With encouragement from Edmunds, he must have felt that the tiresome journey to Manchester might just be worthwhile.

At that stage the luck element still held; the men might not have liked one another; the car might not have been to Rolls's taste. Fortunately, the forthright engineer and the old-Etonian reached an almost instant rapport—and when Rolls saw and drove the 10 hp Royce he knew that his journey had indeed been worth making. At once Claude Johnson, who had been in partnership with Rolls since the end of 1903, was alerted to the fact that the search for a suitable English car was over. Soon afterwards an agreement between the two concerns was drawn up, and well before the end of 1904 a new English car, the Rolls-Royce, was in production. It was also in the public eye, in France: characteristically Claude Johnson had managed to secure space at the Paris Salon, the world's most important motor exhibition.

Since the memorable May meeting a great deal had happened. Although Royce already had a marketable product he was only too anxious to listen to the views of Rolls and Johnson, whose knowledge of motors, motoring and the motorist was so much greater than his own, and at their suggestion he made certain modifications to his original design. He agreed that to concentrate all effort on a 2-cylinder model that might go out of fashion, sooner or later, would be unwise, and readily accepted the idea that a range of models, with engines of 2-, 3-, 4- and 6-cylinders, respectively, should be put in hand [51]. An important feature was to be the use of standard components as much as possible, which simplified production and made significant economies at the assembly stage and in storekeeping.

While Royce and a greatly overworked team at Cooke Street, Manchester, were getting production under way, Rolls and Johnson were making sure of effective

51 *Henry Royce's tidiness of mind was reflected in all his work, as this view of a 1906 30 hp, 6-cylinder engine proves. Its appearance, one feels, was the result of much enjoyable creative work on the part of its great designer and his draughtsmen on the one hand, and, on the other, of those skilled work people who shaped its parts in three-dimensions and put them together with such exactness. Many of its components were common to the sister 2,3 and 4-cylinder Rolls-Royce cars of the 1904–1906 period.*

52 *Although the driving compartment of the 1905, 3-cylinder Rolls-Royce looks crowded to the modern eye, it was laid out with a sense of order rare in its day. Points of interest are the circular ignition control on the steering column; the drip-feed lubricator just ahead of the latter, with its row of sight-glasses showing the driver that oil was on its way to the engine; the large pedals; and the Bowden speedometer shaped like a ship's ventilator, with the total-distance and trip-mileage recorders set transversely, below the round dial.*

publicity for the new car. There was little that they did not know about promotion and sales methods, especially in the quality car market. While it has to be said that his aristocratic background gave Rolls certain automatic advantages in Edwardian society he was no dilettante but an extremely hardworking individual, whose practical abilities as an engineer, a competitive driver, a demonstrator and a motor agent were widely respected. Claude Johnson was also enormously hardworking; one of the greatest organisers and publicists the world motor industry will ever know. Where Edge was noisy and annoying, Johnson was quiet and effective. Each had something worthwhile to promote but Johnson's methods were more subtle inventive and productive in the end. He had the eye for detail which distinguishes the best kind of organiser, a case in point being his thoughtful provision of chamber-pots, discreetly but strategically placed, for the formal opening of the company's new Derby works in 1908. But his overall achievements were on a higher plane than this and without him, his energy and imagination, his endless promotion and protection of his firm's interests and a quite remarkable conception of service, Rolls and Royce, for all their special qualities, might never have succeeded as a duo. Rolls and Johnson were present in Paris, in December 1904 to make sure that the new make got off to a proper start. Outside the Grand Palais they had a 2-cylinder demonstrator, in much demand; inside, on the stand of C. S. Rolls & Co, stood another 2-cylinder (a polished 'show chassis') 3-cylinder, 15 hp chassis, a 4-cylinder, 20 hp car (possibly a mock-up) and a 6-cylinder 30 hp engine. It was an impressive achievement [52].

LATE
EARLY
IGNITION
ONDON &
MANCH
TER

53 *For the first twenty years of motor-car development, most designers who sought more power built bigger engines. The French firm of De Dion Bouton was not after the last horse-power, instead, it concentrated on the production of small, high-speed engines, the durability of which has contradicted contemporary opinion. The number of survivors in this country alone is very high. This 1902 model, with its distinctive bonnet and tubular steel chassis, coped easily with a full load at the start of the 1952 London–Brighton run (above, left).*

54 *Another foreign car, popular in Britain from early in the century, was the Adler. Morgan & Co, of London, handled the make here and supplied most of the bodies for British customers. This 1911 car has a neat landaulette body, with space for luggage on the roof, inside the brass 'fence' (above, right).*

In spite of the success of the Mercedes, French cars were still held in high regard internationally. On road and track the big German machines had the edge—their speed, their acceleration, their relative quietness, their mechanical quality and their visual presence gave them a great advantage—but Delaunay-Belleville, Panhard-Levassor, De Dietrich, Mors, Peugeot and others still sold well, and it was to France, specifically to Paris, that the world looked for the most progressive and exciting ideas in fashion: And what impressive-sounding names some of those French coachbuilders had—Labourdette, Rothschild et Fils, Milion et Guiet or P. Charpentier et Cie. The superiority of the Mercedes and of certain French makes was not seriously challenged by any British car outside these islands.

In Britain sales of foreign cars were comparatively small but sustained. What must be called lesser foreign makes, the French Renault and De Dion [53] in particular, and Adler [54] and Opel from Germany had a steady sale. From Italy FIAT [55] and Lancia had begun to find a small but faithful following and American sales were already quite substantial.

The combination of Rolls, Royce and Johnson was soon to change matters; as far as cars of the highest quality were concerned the 40/50 hp Rolls-Royce which appeared in 1906 and was in production by 1907, took a convincing lead [56]. It

55 *Italian cars have sold well in Britain since the early 1900s. This* circa *1908 FIAT model has three rows of well-upholstered seats in its large Roi-des-Belges-type body, with a narrow passage between those in the centre to allow access to the high-seat rear compartment. Its rear tyres are steel-studded, the side door has a security latch above the handle, its horn bulb is set in the centre of the steering column and a bonnet strap gives a somewhat sporting look to what is, in fact, a luxurious touring car.*

56 *The thirteenth 40/50 hp Rolls-Royce chassis (the only one officially named 'Silver Ghost' by its manufacturers) was fitted with a fine-looking touring body on Roi-des-Belges lines, made by Barker. In 1907 Claude Johnson entered it for a 15,000 mile non-stop reliability trial, organised and observed by the Royal Automobile Club, and its success was a vitally important factor in establishing the Rolls-Royce reputation for reliability and longevity. The tulip-shaped rear of the body of this famous car is characteristic of the type. Other features include the tool-locker immediately above the chassis cross-member, the platform-type rear suspension, the broad, flat mudguards and the large speedometer, with separate distance recorder above, set outside the body, near the horn bulb.*

was quite simply the best machine of its kind in the world. To find a new selection and arrangement of words to describe it adequately is too hard a task and no apology is made for borrowing a description of an 'improved steam engine', designed by Trevithick and Vivian close on two hundred years ago. The writer was the 'eloquent Mickleham' and he might have been describing the 40/50.

'It exhibits in construction the most beautiful simplicity of parts, the most sagacious selection of appropriate forms, their most convenient and effective arrangement and connexion; uniting strength with elegance; the necessary solidity with the greatest portability; possessing unlimited power with a wonderful pliancy to accommodate it to a varying resistance.' Its Manchester-built predecessors, the 10, 15 and 20 hp models, and the contemporary 30 hp 6-cylinder, had built up a remarkable reputation in two years for refinement, reliability and resistance to most of the normal processes of wear. The success of the Light Twenty in the 1905 and 1906 Isle of Man Tourist Trophy road races had given the young company useful publicity and helped sales, morever, with a certain number of cars being sold abroad. After the 1906 victory, Rolls took the winning car to America, where he had some competition success before finding a buyer, and one or two 30 hp cars were subsequently sold there. But it was Royce's masterpiece, his largest car yet, that put Britain firmly in the lead. It was handsome as a chassis, which encouraged the coachbuilders to greatest effective effort, Barker [57], Hooper and H. J. Mulliner

57 *Baron Korff ordered this 40/50 hp Rolls-Royce in 1910 or early 1911, at the time when the ample curves of the Roi-des-Belges and similar types of body were giving way to plainer, flatter shapes. Its splendid Barker tourer has a Cape Cart hood, with straps at the front to secure it to the chassis, electric side-and-tail-lamps, a glass-topped scuttle to throw daylight on instruments and controls mounted on its inner side, an elaborately hinged windscreen, with a separate one for rear-seat passengers, and a small luggage grid at the back of the body.*

setting the pace for all the others. At first, prevailing fashions being what they were, styles were somewhat clumsy; shapes and proportions uneasy. The closed bodies, on the whole, were the least satisfactory in appearance, but matters improved and by 1910 the adjective 'handsome' could be used without qualification in numerous cases. When everything was right there was nothing quite like a good 40/50. It was a near-perfect fusion of the best available high-power chassis and the best possible bodywork ([58] shows the high standard reached early in its production life), comparable with man's greatest achievements in other fields of transport—the windjammer, for example.

Most important, the 40/50 Rolls-Royce quickly established an international reputation for excellence that reflected favourably on British engineering as a whole. The combination of a distinctive and elegant appearance with tremendous reserves of strength, and a highly favourable power-to-weight ratio gave it a particular appeal to the wealthy and discerning in countries where driving conditions were difficult. This was so in India, for example, where there were many eager customers, after a specially-prepared 40/50, 'Pearl of the East', had been demonstrated there, in 1908. The connection then begun between Eastern rulers and Rolls-Royce was to be a close and lasting one. Other British cars of quality also sold well in India: Daimlers and Napiers, for example, and one Indian Prince, Ranjitsinhji, a world-famous cricketer, had so strong and enthusiastic a liking for Lanchesters that he owned something like

58 *Barker & Co. were always expert at producing formal coachwork that was above average in appearance and finish. Although this 1910 40/50 hp Rolls-Royce is at least seven feet high its great length and the strong horizontal emphasis, as well as excellent proportions, disguise the fact. The straight-sided bonnet is characteristic, as are the brass rails surrounding the luggage space on the roof, the speaking tube placed close to the chauffeur's right ear, the railway-carriage-type door handle and the visually satisfying contrast between straight lines and curves.*

59 *Rolls-Royce learned a great deal from intensive development testing abroad and by active involvement in the Alpine Trials of 1912, 1913 and 1914. Engine-cooling was improved with larger radiator area and well-louvred bonnet sides but the special steam 'dome', seen here on the 1913 Trial-winning car, was not used for later production cars. It served as a steam condenser. The appearance of this large touring car is superb. It set the fashion for sports and touring bodies during the 1920s (above).*

60 *The neater shape of the motor-car was the result of a continuous process of development that is still underway. The 1913 Rover Twelve tourer, with its maker's own bodywork, was abreast of contemporary fashion, with flush sides, tidily folded hood, comfortable diamond-pleated leather seats, folding windscreen and rounded scuttle. Electric lighting was an optional extra (below, left).*

61 *The original Rover Eight of 1904, was of outwardly conventional appearance; a lightweight runabout of great simplicity that was easy to build, cheap to run and simple to maintain. Mechanically speaking it was far from conventional, with its 'backbone'-type frame, steering column gearchange and pedal-operated parking brake; the latter was to be a feature of Citroën DS19 cars fifty years later (below, right).*

forty different examples. Most of the cars going East had bodies of flamboyant appearance and colour, and were richly and expensively furnished within. They had something of the spectacular presence of the more impressive kinds of carriage. Indeed, carriage bodies were fitted to some, and rather odd they looked: the English coachbuilders had been well established in these markets even before the coming of the automobile.

The Rolls-Royce was a British car suitable for worldwide use, with little modification—well-louvred bonnet sides were available for colonial service, and a slightly enlarged radiator [59]. Other makers produced special colonial models, which normally featured lower gear-ratios and larger-diameter wheels and tyres. Napier and Rover produced typical examples on their 15 and 12 hp chassis respectively.

The Rover Twelve [60] was one of the best touring cars of its time. Its makers had diversified into car production in 1904 after establishing a high reputation in the cycle and motor-cycle industries. Their first model, the Eight [61], had been designed for them by a former Daimler man, Edmund Lewis. Externally it looked conventional, but its frame consisted of a single, tubular aluminium structure, joining engine, clutch and gearbox to the rear axle casing, which was unconventional for its day. Very much later this plan was to be adopted and suitably adapted by Ferdinand Porsche and others.

The Eight, which cost £200, and the Six, which was added to the range in 1905 and cost £105, were very successful and remained in production for many years. Other, larger Rover models were reasonably popular and one, a slightly modified 16/20 tourer, attracted useful prestige and publicity for its makers by winning the 1907 Tourist Trophy race (by luck, more than anything, it must be said). It was the Twelve that Owen Clegg, formerly of Wolseley, designed soon after joining the Rover company in 1911 which set the make above so many of its contemporaries. On paper there was nothing out of the ordinary about its specification but Clegg was an extremely intelligent man, who had made a detailed study of market requirements, and he planned a car that could be produced economically without any sacrifice of quality. It had a neat-looking 4-cylinder engine, a three-speed gearbox and final-drive by underslung worm. Its Rover-built touring bodywork was handsome and well made; the chassis price, with tyres and spares included, was £275, the 'Finished Car', with five-seater body cost £315, and the 'Complete Car', with hood, screen, two acetylene headlamps, two side lamps and one tail (oil or electric at purchaser's option), horn, spare wheel and steel-studded tyre, was very good value at £350. It was the 'Car of the Year' as far as 1912 was concerned, sold widely and well and remained in production until 1924, by which time it was known as the Fourteen. Its engine had always been of 13.9 hp on the RAC formula (at the time of its introduction, a size more common abroad than here) but it developed 22 bhp in its original form and a shade over 40 bhp in its last. As with so many first-rate touring cars of its type and period, however, the great attraction of the Rover was its manner of going. It had a very quiet and relaxed engine that was still unfussed at 40 mph; a precise and light-acting gearchange and its steering combined accuracy and directness with lack of effort.

62 *The Calcott was not made in large numbers and there was nothing exceptional about its design. It was, however, an excellent miniature car; simple and reliable and very pleasant to drive. Mr Lumsden's 1914 model has been a familiar sight on the road in Scotland and elsewhere for the best part of twenty years, a veteran car used and enjoyed for its own sake. Its two-seat body probably cost well under £20 to build (above, left).*

63 *The Scottish-built Arrol-Johnston dogcart, in the six-seat, 12 hp version, seen here outside John O'Groats hotel at the start of a successful drive to Land's End, towered above most of its contemporaries. Its tall, mahogany-panelled body was made by a celebrated Glasgow cabinetmaking firm and was notable for its high-varnish finish. Weather protection was not thought to be important. (above, right).*

Other contemporaries that reached a comparable standard, or exceeded it, were the excellent Sunbeam Twelve and Sixteen, the 15 hp Napier and the Wolseley 16 and 20 hp models. The largest Daimlers, Sunbeams and Wolseleys, too, ran with the special ease of a good 6-cylinder car and were most satisfying to drive. Those accustomed to the modern motor-car might not be impressed at first by engines rather more audible and certainly less responsive than those in present use, but on the other hand, he would quickly appreciate their strong low-speed pulling powers, flexibility and lack of mechanical strain. With some difficulty, it is true, he might master the problems of changing gear without synchromesh, but if he succeeded he would discover the sensual satisfaction of effortlessly swinging a hefty right-hand lever 'across the gate' into the chosen gear without a sound. The high-geared steering of the majority of these cars was another major attraction; very direct, so that well under two turns of the steering wheel took the car from lock to lock, with absolute accuracy and surprising lightness.

It must be said that many early cars are thoroughly unpleasing to drive. In their day such things as unresponsive engines over-sensitive to carburetter and ignition settings, slow, difficult and noisy gearchanges, too-lively or completely dead steering, and ineffective brakes were tolerated because English motorists knew no better. The differences between a 6 hp Rover, say, and an 8 hp Renault twin are very marked,

64 *High finish is a feature of this 1898 Daimler dogcart with hood, the body of which is so well-proportioned and possessed of so many subtleties of shape and decorative detail that it is almost certainly a Mulliner production. It is a specimen example of the horseless carriage.*

65 *During the ten years that separate the Daimler in 64 and this 1908 Daimler (disguised for an unknown reason with a different kind of radiator, bonnet and scuttle), the motor-car developed considerably in appearance and in mechanical effectiveness. Daimler bodies were named after English cities, towns and villages. This tourer, with its Cape Cart hood and mudflaps behind the front wheels, is a 'Highgate'.*

the French car willing, responsive and in no way demanding to drive. The Rover, on the other hand, is tiresome; far too demanding of its driver's attention. One of the most wholly agreeable light-cars within the writer's driving experience is the 1914 Calcott [62]. It was among the first, and best, of the miniature models to appear in Britain from 1912 onwards. On suitable roads, to this very day, long journeys can be covered at a steady 30–35 mph with complete regularity and reliability, and with vast enjoyment. Fortunately the best characteristics of these so-called Edwardians were carried forward into the 'twenties.

The gradual alteration in the appearance of the motor-car between 1900 and 1914 has already been noted; the later types, open and closed, having a simpler, neater and more unified shape overall that was to be further modified and improved in many ways during the next decade or so. The oddities had long since gone; no longer did the tall, archaic-looking Albion or Arrol-Johnston dogcarts [63] scamper about Scotland, the country of their origin, on high, wood-spoked and solid-tyred artillery wheels, carrying doctors on their rounds. The Sunbeam-Mabley had disappeared, unlamented; the Orient Buckboard was no more; the Lagonda and Rover tri-cars too, had passed from grace. The early, top-heavy looking Daimlers had been replaced by more handsome cars [64, 65 and 66]. Excellent as the current Lanchesters were, in 1914, they lacked the individuality of the early twin-cylinder

66 *A year or two later this large Daimler was on the road, with a distinctive boat-shaped body somewhat ahead of its time. The junction with the long, flat-topped bonnet is not a happy one. Bodies of the type were popular between 1912, say, and 1914, and in the early post-war period.*

67 *For a year or two before the beginning of the First World War a certain amount of effort was expended in lowering the wind resistance of cars specially prepared for record breaking or racing. Among those interested was Herbert Austin, who took the matter very seriously, with this extremely narrow single-seater. Note its cowled radiator, disc-wheels, faired-in dumbirons and wind-cutting axle. Marine and, perhaps, aero influences are discernible.*

cars, but their greater orthodoxy made them acceptable in market terms. There was a renewal of interest in easing the passage of the automobile through the air by cleaning up its shape as much as possible [67].

The gauche appearance of the Scottish challenge to De Dion and other makers of the *voiturette*, the diminutive Argylls of 1900 and 1901, had disappeared. The light-car of 1914 was no compromise at all; it was well considered, well-made and well bodied. Apart from Calcott, other good makes were Austin, Calthorpe, Humber, Lagonda, Singer and Standard. The 1,100 cc Lagonda that appeared in 1913 had unit construction of body and frame, of a rather rudimentary kind. Larger models of that make sold well to Russia.

The most interesting of the small cars of the period was the Morris Oxford [68] that went into production in 1913 and was the first serious and successful example of the assembled car to be produced in Britain. William Morris had a wide experience in the motor trade and made a deliberate decision to create a high-grade small car that would be reasonably priced, reliable in use and inexpensive to run and maintain. He went to specialists for his engine, gearbox, final-drive units and so on. They were well-chosen; the balance of qualities was a good and attractive one. The little Morris was speedy, reliable (as long as the correct sparking plugs were used) and likeable, and it was an immediate success. In 1914 it was still small fry on the British market. Within ten years it was the best-selling car in Britain.

The Oxford appeared in prototype in 1912, when the cyclecar was under intensive development as an alternative to the motor-cycle and sidecar on the one hand and the smallest, cheapest and nastiest kind of car on the other. Most of those concerned with its design and manufacture, however, seemed to be more interested in reducing prices than in raising levels of quality and reliability. In competition with motor-cycles of real excellence, such as Humber, Rover or Sunbeam, for example, or small cars as good as the 10 hp Calthorpe, the new Morris, the smallest Singer or the

68 *The Morris-Oxford was an immediate commercial success in 1913. It was an attractive looking light car that provided quiet, comfortable and reliable transport for two. When this 'bullnose' type went out of production in 1926 its basic appearance had changed little.*

69 *The idea of producing a small, simple and inexpensive vehicle, combining car comfort with motorcycle performance and economy, to cater for what might now be called the marginal motorist, was sound enough. In practice, poor design and low standards of construction made the typical cyclecar more of a mechanical nuisance than anything else. Tandem seating was one feature of this 1913 French-made Super Cyclecar. Unreliability was another.*

70 *The Oakland was an American car seldom seen in Britain either before or after the First World War and it was not a leading make in its own country. Nevertheless, in 1914, someone commissioned the Grosvenor Carriage Co. of Kilburn, London, to build this shapely coupé with folding head, comfortably upholstered 'dickey' seat at the rear, electric lighting and detachable-rim wheels, the latter fitted with Goodrich tyres. It is visible proof of the high standard of design, build and finish that the smaller London coachbuilder could achieve.*

9.5 hp Standard, cheapness was simply not enough for success. Some cyclecars had the look of a bath on wheels; the French Super Cyclecar [69] is of that type. Others seemed to be powered, larger versions of childrens' pedal cars. As we shall see in Chapter 4, an attempt to revive the cyclecar immediately after the First World War failed because of the superior attractions of good small cars like the Rover Eight and the Austin Seven.

In 1914 the British motor, motor accessory, tyre and coachbuilding industries were in a strong and prosperous condition. In many important respects they were ahead of foreign competitors. Generally speaking, however, the majority of manufacturers were producing vehicles best-suited to home conditions and needs, and from that point of view they were to find themselves at a disadvantage at a later date. Meanwhile they went on turning out an increasing number of private car, taxi, lorry and bus chassis, with bodywork to suit, at all price levels. The reputation of the leading London and provincial coachbuilders was deservedly high, and in certain instances it was international. The work of lesser concerns reached more than acceptable standards of design and construction; plate 70 shows a good example of Anglo-American collaboration at this level. As we have already learned, certain car manufacturers ran excellent body departments of their own. The 1914 Austin illustrated in plate 71 is proof of the competence of the Longbridge design and bodybuilding staff.

71 *Any enthusiastic motorist who has driven one of the better cars built in Britain between, say, 1912 and 1914, knows how satisfying such an experience can be. It is easy to imagine the ease of running of this large Austin (almost certainly a 1914, 50 hp model) the pleasures of control and the sheer physical delight of being in such a handsome machine. Its long bonnet houses a 6-cylinder engine and its beautifully proportioned body, with inviting diamond-pleated upholstery seats five, perhaps seven, for there is space for occasional seats at the rear. The fact that the lighting is by oil (for the side and tail-lamps) and acetylene (for the large headlamps) is not unusual; electric lighting did not come into widespread use until the 'twenties.*

3 *The First World War* 1914 *to* 1918

At the beginning of the war, in August 1914, the motor vehicle was an accepted fact of everyday life in Britain. Ownership of private cars was largely limited to the better-off, as was use of taxi-cabs, but the gradual spread of petrol-driven buses was familiarising more and more people with the convenience and great operational flexibility of the new vehicles. The greatest growth of these buses was in the London area, where their original purpose had been to serve as feeders for existing tram and suburban train facilities. In other cities and towns the electric tramcar still prevailed [72]. With notably improved reliability the value of the motor bus for more general use could be exploited at last. A strong point in its favour was the fact that it could run on existing roads, without specially laid track. Buses were also to prove useful in active service, large numbers of London General Omnibus Company double-deckers providing reliable troop transport in France in the early days of war, carrying more soldiers at a single loading than the Paris taxis, impressed by the French Army for a similar duty.

The Western Front was no place for the mechanical weakling. The going was awful for most of the time; the mud omnipresent, with shells and shell-holes an unwelcome extra hazard. The horse and mule were much used by the military, and for certain tasks they were invaluable; but the overall superiority of the motor vehicle was proved over and over again. The lorry, in particular, came into its own, on both the Allied and the German sides.

In 1912 the War Office organised its Subsidised Vehicle Trials as the first move in a scheme for the creation of a pool of load carriers of approved types to be available in emergency. The basic idea was that buyers would receive a purchase premium of £50 and an annual subsidy of £20 for three years as long as the vehicles were properly maintained. If war began, owners had to hand them over, on demand, within seventy-two hours. It was a sensible scheme, and one that guaranteed an immediate supply of suitable lorries. France and Germany had similar arrangements, but offered better terms to those taking part.

It was not only goods vehicles and buses that were commandeered for immediate military use. The capacity of the 40/50 hp Rolls-Royce to withstand the hardest driving, over long distances and on any kind of road, had been strikingly demonstrated by its successes in the 1913 and 1914 Austrian Alpine Trials, and this had not passed unnoticed by certain motor-minded officers. Since 1912 Rolls-Royce had been competing in this very demanding event, and had learned much of practical value in a short space of time. James Radley's much-publicised failure during the 1912 Trial to make a restart on a one-in-four gradient until he had cast out a couple of passengers, had important consequences; the three-speed gearbox of the 40/50 was replaced by a four-speed unit, and minor changes were made to the carburation. From more general experience, a certain amount of strengthening of the chassis was carried out, as well as improvements to the cooling system. Most important of all, however, was the establishment of a programme of high-speed, long-distance development-testing on the Continent which continued until the outbreak of the next war, in 1939. One British manufacturer, at least, was determined that his products should be suitable for worldwide use.

It was the known combination of reliability with extreme strength that attracted military minds, and armoured-car versions of the 40/50 were built with amazing speed, once war had been declared. As part of the RNAS, they were sent out to Belgium as soon as they were ready, to engage the highly mobile German forces (mainly cavalry) and their behaviour in action was so satisfactory that further squadrons were formed for service in other war areas. Even weighted down with tons of armour-plate, arms, ammunition and crew the big Rolls-Royces gave a spirited performance on suitable surfaces. Their quietness often gave crews the tactical

72 *In 1914 a set of photographs of traffic conditions in Glasgow was made. At that time, city streets were choked by tramcars and horse-drawn vehicles of all sizes and kinds, motor vehicles being few in number. In this view of rush-hour congestion on Jamaica Street Bridge, over the River Clyde, only three cars are visible. There are ten tramcars.*

advantage of surprise. Operation of the cut-out which allowed exhaust gases to escape directly to the air, gave an extra spurt of speed that was equally operationally advantageous on occasion.

Rolls-Royces were also used to transport Generals and others considered sufficiently important [73]. Three specially selected cars covered enormous distances during the war on the King's Messenger Service which, for centuries, had provided personal couriers to deliver Government intelligence to British ambassadors in foreign lands. The service had formerly been carried out by 'retired military men or naval officers, or other hardy and adventurous gentlemen'. They had made their slow journeys across Europe in specially-equipped carriages of what would nowadays be called a heavy-duty type and a good messenger was expected to deal with minor repairs *en route*, a well-stocked tool chest and a supply of spare parts being aboard his conveyance.

Of course, Rolls-Royce was not the only make claimed for Services or official use. Crossley found favour with the Royal Flying Corps (later the Royal Air Force) as staff car, tender or ambulance. Daimlers were used as staff cars by the Army; because their engines were of the Knight sleeve-valve type, starting problems during continental winters were not unknown. Lanchester made searchlight-tenders and field kitchens as well as a number of armoured-cars, all on modified 38 hp chassis. Several squadrons of the latter proved effective over great areas of the Russian front, where conditions could be dreadful. Among other specially-selected British cars were the 16 hp Sunbeam, numbers of which were built by the Rover company, the Talbot, the 25 hp Vauxhall—a particularly sturdy and popular model—and the Wolseley.

73 *This circa 1911 Rolls-Royce 40/50 was built for peaceful purposes but its inherent strength, power and already legendary reliability made it eminently suitable for military duties. Whether carrying high-ranking officers and politicians about their wartime duties or serving as a fighting vehicle, laden with tons of armour and weapons, these immortal cars played a significant role on many fronts and in many lands.*

VIII *The Cooper organisation and other similar concerns quickly adapted to the specialist body design requirements of the motor industry—working drawings, 'visuals' and colour suggestions being provided from an early stage. Sometimes the ideas proposed seemed to verge upon the preposterous, this design for a bus,* circa *1912, being an amusing example. It was, however, a practical proposition technically, and its appearance might have been a commercially valuable feature.*

IX *Body changing of the kind proposed in this drawing was no novelty in the early days. With a separate chassis it was not at all difficult to effect, and the practical result was greatly increased use of the vehicle in question. It was not uncommon for a lorry to carry coal during the week and passengers (in a suitable alternative body, of course) at weekends.*

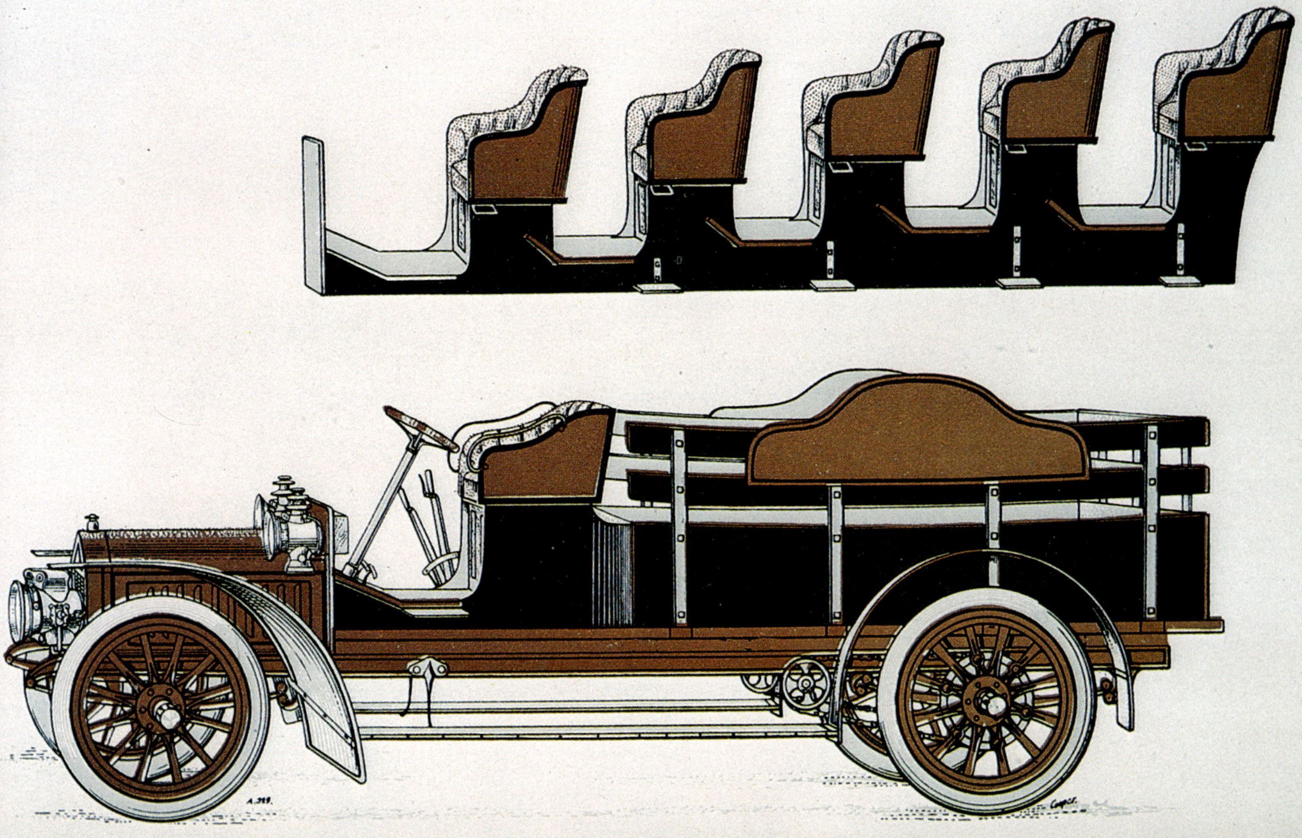

X *This design for a sporting-car,* circa *1914, is an excellent example of the way in which the shape of the car was simplified in little more than a decade.*

XI *The 45 hp Renault was the largest model in an extensive range of private cars. In the early 1920s its makers distributed this well-sculpted, skiff-bodied miniature to their agents, some using it as a paperweight; others, it is hoped, valuing it for its aesthetic merit.*

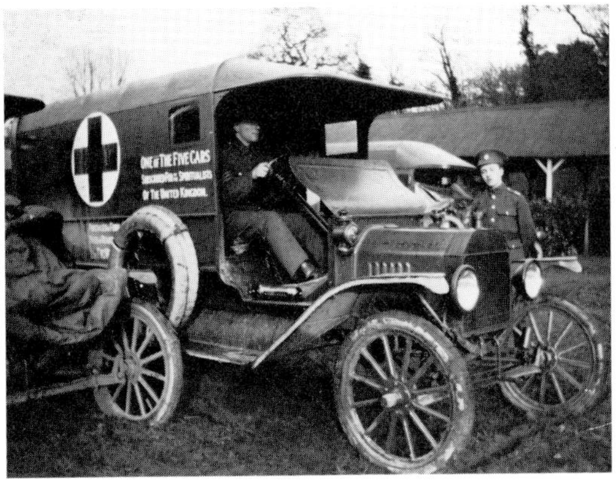

74 *Vast numbers of Model 'T' Fords were used during the First World War, their great mobility and basic reliability making them particularly suitable for off-road work. This ambulance was one of five subscribed for by the 'Spiritualists of the United Kingdom'.*

75 *The unknown artist responsible for this commemorative plaque (presented to Halley Industrial Motors by a grateful employee) knew his subject intimately: no detail of its special equipment has been omitted—sturdy bodywork, towing hooks at front and rear, canvas hood, the row of petrol cans along the cab roof, and so on. As automotive art, it ranks highly.*

One car that looked as if it might not stand up for long to wartime conditions was the Model 'T' Ford [74]. It was, in fact, an enormous success, a major advantage being the combination of high torque, or pulling power, with low weight. A Ford tender or ambulance could often carry on when other vehicles of higher power, and weight, became bogged down. The Ford two-point suspension also gave it an extraordinary cross-country performance. It was the ancestor of the Second World War jeep, of the later Land-Rover and other specialised vehicles; like them it could dart across rough ground with the alacrity of a questing ant.

Such mobility was beyond the powers of the large, heavy lorries being used in increasing numbers at, and behind, the various fronts. The fact that the majority ran on solid tyres had much to do with this. The French Michelin company had introduced pneumatics for goods vehicles in 1912 but they had not caught on in Britain, and, indeed, they were not to do so until the mid-'twenties. Even less manageable in the mud were the steam lorries, of which there were many, Foden alone having supplied something like a thousand.

The petrol engine was so much easier, so much more convenient to deal with, and starting it was not to be compared with the long-drawn-out ritual of raising steam. Approved types of vehicle were built on a large scale. Upwards of 10,000 AECs saw service. There were 7,500 Dennis lorries and a similar number of Wolseleys. Leyland and Thornycroft each supplied more than 5,000 vehicles. The Leyland was the longest lasting of these, some still being in civilian use after the Second World War. The Army Service Corps received about 250, the Indian Army half that number, and about 150 were released for civilian use. The largest user was the Air Force, which had more than 5,500 in all, operating as heavy tenders, mobile workshops and special-purpose vehicles. Other makes included Albion (then the principal Scottish manufacturer) and Halley [75], Caledon (also Scottish), Austin, Burford, Hallford, Karrier, Maudslay, Napier and Star.

76 *In Britain one method of saving scarce petrol was to use alternative fuels. Town gas was a favourite, and the balloon-like bag secured to the roof of this Scottish Motor Traction Lothian bus is its wartime fuel container.*

Not all the lorries survived the war, naturally, but so many did that their disposal became a major problem after the Armistice. Leyland, a maker with a conscience, it would seem, bought back as many survivors as possible for proper reconditioning before resale on the civilian market. Altogether more than 3,000 vehicles were dealt with at their Kingston-on-Thames factory. Nor was all the necessary work of a mechanical kind; canvas was all very well as a cab-roof material for military vehicles but something more substantial was thought necessary for civilian users. Moreover much wartime lorry bodywork was of highly specialised type and had to be extensively modified, or even scrapped and replaced with something more appropriate. It was less easy to dispose of the special-duty vehicles—ambulances, field-kitchens, gun-limbers, ammunition or stores transports, mobile workshops and canteens among them—that were built to meet military needs by the coachbuilders of Britain and made to be drawn by horses, mules, camels, oxen or even steam tractors. Animal power, it must be emphasised, did have advantages under certain conditions of modern warfare, and it was only the cavalry charge that had gone completely out of date.

The arrival of war caught most people by surprise although the possibility had been discussed for many years. In some respects life went on normally—in others wartime shortages manifested themselves [76]. Many 1915 cars were introduced by the motor industry in the usual way and manufacture continued. By 1914, motor vehicle appearance had reached a state of some stability and 'next season's' models differed little from their predecessors. It was a matter of gilding existing lilies rather than producing fresh blooms [77]. However, the war was not over by Christmas, as so many had predicted. The making of munitions became a great deal more important than the manufacture of motor-cars for civilian use, which had ceased, more or less, by the end of 1915. There was one exception: William Morris had secured contracts for the production of mine sinkers and other warlike stores, but he had also been receiving supplies of components for his Cowley model from the United States during 1915, and as long as full sets were available he continued with their assembly. Of the 3,000 American Continental engines, ordered during his visit to America in August 1914, some 1,500 were lost in the Atlantic; by the Armistice, inevitably, he had a reserve stock of other parts.

Morris was a patriot, but he was also a businessman and a realist. To meet American competition in his own country he was compelled to take advantage of the facilities available in Detroit for the quantity manufacture of components, at prices far below those prevailing in Britain. White & Poppe of Coventry, who were specialist engine manufacturers, and who already supplied the engine for the Oxford, quoted £50 per unit for the new Cowley. The Continental Motor Manufacturing Company of Detroit, quoted £18. Differences of that order made it economic sense to ship American-designed and American-made parts to England for assembly; until the imposition of the thirty-three and a third per cent McKenna import duty on certain 'luxury' goods, in September 1915, and the subsequent ban on the importation of American components, except for goods-vehicle manufacture, began to cancel out the advantages.

The Cowley was an early example of successful two-nation collaboration, and the second Anglo-American enterprise on such a scale, Ford being the first of this kind. It was very much an English car, in fact, despite its largely transatlantic content, and one of those rare models that was successful from its very beginnings. There was no

77 *Dr Low, the uniformed driver of this wartime Hillman, which might well be of much later date with such clean and pleasing lines, later became a prolific contributor to the British motoring press. His car has auxiliary springs attached to its rear shackles; a shapely, rounded tail, and a generally neat appearance.*

question at any time of direct American involvement until the mid-'twenties, firstly with the setting-up of the Pressed Steel Company and secondly with the acquisition of the financially ailing Vauxhall concern by General Motors. The British industry then became more closely connected with Detroit.

American automobile production continued, uninterrupted, because of that nation's delayed entry into the war. The most significant development was the appearance of a method for producing all-steel bodies in quantity, and not before time, as the expanding demand for automobiles was placing impossible demands on the bodybuilders. That they were able to cope successfully for so long had something to do with the fact that even in an industry as labour-intensive as car manufacture there had been no real shortage of workmen. Henry Ford's magicians, his production engineers, had shown what could be accomplished, but obviously the last stage of the production process, the making, trimming, painting and finishing of the bodywork, could be speeded-up a great deal.

For some time past, steel panels had been in general use in America and, to a lesser extent, in Europe, as external cladding over wood framing, for bodywork produced in quantity; mahogany or, more commonly, aluminium remained the preferred material for the specialist coachbuilders. Discussions took place in 1913, between the Budd Manufacturing Company of Philadelphia, and Dodge Brothers about the supply of steel panels for their forthcoming car and E. G. Budd was able to convince John and Horace Dodge that it was possible to produce an all-steel body. They authorised him to do so and the long and difficult process of design, die-making and press-installation began. By 1916 an all-steel touring body was being stamped out in substantial numbers and it was soon followed by a saloon, notable for the subtlety of its curves and its large window areas. From then on the days of composite construction were numbered, as far as volume-production bodywork was concerned.

Although car manufacture was greatly reduced in European countries and there had to be a technical standstill, enormous advances in aero-engine design were being made. Before the war the race tracks and circuits had been the forcing-houses, with France and Germany tussling for superiority. Now it was the air.

The aeroplane, which had been taken less seriously than it deserved by the British Army and Navy, soon acquired great military importance and, inevitably, the motor industries of the combatant nations became deeply involved in its manufacture. Aero-engine design and development was a vitally urgent matter, and large-scale production a major priority. Some firms made themselves entirely responsible for the manufacture of aero-engines, from design to completion; others built to outside designs. Humber, for example, made the W. O. Bentley-designed BR rotary, which was a 9-cylinder, air-cooled unit of 250 bhp, and Wolseley produced the splendid-looking Hispano-Suiza 90° V-8, the design of which was to influence certain of their own engines once the war was over. Napier's own 450 bhp, 12-cylinder, 'broad-arrow' formation 'Lion' was a masterpiece of its kind, and the first of a range of outstanding aero-engines. It was to power aeroplanes, high-speed launches and the occasional land-speed record-breaking car in later years. Rolls-Royce, according to an anonymous writer in 1919, had 'the suffrages of all pilots'; certainly its series of engines, all water-cooled, established an unequalled reputation for reliability and high power output. The 'Eagle' was built by Straker-Squire, the Bristol car firm, and before the end of the war plans were at an advanced stage for large-scale production of components in the United States. Although this project came to nothing because the war ended, it led, before long, to the foundation of Rolls-Royce of America Inc.

This was a brave attempt to take the sales war into the enemy's camp, as it were. The 40/50 had sold well in the States before August 1914, but afterwards, when cost levels were at least double those prevailing in 1914 and an unfavourable import duty prevailed, it seemed that the selling price would be too high, even for the millionaires. It was known that on-the-spot manufacture of a chassis identical in every respect to that produced in Derby, and made to the same levels of excellence, could keep the cost to a reasonable figure, and it was hoped that this would stimulate greater demand. In theory the scheme was sound. In practice, however, it ran into various problems arising from different attitudes towards car purchase, ownership and use. It was the third significant example of Anglo-American collaboration in the automobile industry. It nearly succeeded.

4 *The vintage period* 1919 *to* 1930

Amidst the general euphoria of post-Armistice Britain the making and marketing of motor vehicles seemed to offer almost limitless commercial possibilities. During four tedious, terrible years of war, a large demand from established users had built up, an altogether new demand had grown among the multitude who had been introduced to the motor vehicle by one or other of the Services, and had come to relish the promise of freedom and mobility for the individual once normal living could be resumed.

It is scarcely surprising, therefore, that so many optimists sought to take what share they might of a market for transport greatly in excess of anything that had previously been considered possible. The demand was not for motor-cars alone. Under the exceptional conditions of wartime use, the flexibility, versatility and reliability of the road vehicle as a carrier of goods and men had been closely observed by the imaginative few, who saw in road transport the possibility of a commercial future. With the return of peace they began to exploit its potential in a variety of ways, greatly helped at first by the availability of ex-Service vehicles of familiar, well-tried types. There was also a ready supply of drivers and mechanics with much valuable practical experience to draw upon.

An especial favourite with owners and drivers alike was the big, bear-like RAF-type Leyland that became the mainstay of so many transport fleets, large and small, throughout the 1920s. Other similar British lorries that were almost as popular and gave equally meritorious service included AEC, Albion, Daimler and Thornycroft. The much smaller Crossley RAF tender, with its distinctive twin-rear wheels, was used for passenger-carrying as well as goods transport. Of the American makes left behind, Packard, Pierce-Arrow, Peerless, Liberty [78] and FWD are particularly remembered. These larger lorries were widely-used for long-distance haulage between the major industrial and commercial centres—a fact that did not please the railway companies.

Their mechanical strength had to be matched by the physical stamina of their crews. Power assistance of any kind was quite unknown, and the amount of effort required to change direction, change gear or operate the brakes in urban driving or on an extended journey in difficult terrain, often with a bulky trailer in tow, would be considered quite unacceptable now. Crews had to be tough, to resist the elements.

78 *The Anglo-American Oil Co. founded its post-war fleet on army surplus vehicles of this kind. It is an American Liberty, a four-tonner that delivered paraffin in the Nottingham area until 1932. The cab doors may have been a later addition.*

79 *After the Armistice it was unusual to find lorries so completely devoid of weather protection. In spite of the minimal bodywork, this Karrier took a load of five and a half tons of Pratts petrol, in two-gallon cans. The 'BY APPOINTMENT' shield is proudly and prominently displayed on the body side.*

Drivers' cabs sometimes lacked a windscreen and doors in the earliest days [79] and heating systems did not then exist, although there was an ever-present spill of engine heat through the bulkhead and pedal slots to provide some relief. In summer time, however, cabin temperatures could soar to almost unbearable levels. Certain car drivers had to endure exposed conditions as well [80]. They also had to be resourceful to deal with the lesser kind of mechanical malfunction, with loads that moved or with tyre trouble. Although 'solids' did not burst, they could break up or fly off wheel rims, and the incidence of pneumatic punctures was high, largely because of prevailing road conditions. Surfaces were much less uniformly good then and nails from horseshoes, for example, were still a widespread hazard. Conditions did improve steadily, however. Windscreens, doors and side windows gradually became standard equipment. The windscreen-wiper and other accessories, such as instruments, made a somewhat tardy appearance. The changeover to pneumatic tyres improved riding comfort and reduced noise levels dramatically, especially on the cobbled or paved thoroughfares then common in many built-up areas.

Vehicle sounds were a good deal more varied and distinctive then. Perhaps the noisiest of all was a solid-tyred, chain-driven lorry (like the Karrier), running light over granite setts or pavé. A well-used bus could come a close second in similar conditions; the rattling of loosened seats and window glasses accompanying squeaks and groans from the body frame, the thunder of tyres and the roar of an unsubdued engine. Here again, however, the rate of change in design brought swift improvements, the change to the pneumatic tyre making the most immediate difference [81]. Quietness was not thought to be of prime importance as far as goods vehicles were concerned (though certain models are remembered in that respect) but it was taken seriously enough by omnibus manufacturers. It is on record how certain Midland Red SOS single-deckers of the late 'twenties slipped quietly along at speeds greatly above the legal limit and how smooth and mechanically unobtrusive were the big Leyland Lion and Tiger buses and their two-storied stablemates, the Titans. The performance of the smallest Commer coaches, around 1930, and of Bedford buses a year or two later, was comparable with that of contemporary medium-size cars, not only in terms of ultimate speed but in running refinement as well. Their engines combined the silky quality of a good six with quiet vigour.

Engines, except those of most sports-cars or the kind of raucous, air-cooled twin favoured for the smaller types of economy runabout, were less obtrusive, often, than the gearboxes to which they were connected. The 24 hp Albion chain-drive lorry emitted a characteristic screech of protest every time its clutch was carelessly operated. Steam vehicles made little mechanical noise; instead they puffed gently, but audibly from time to time, emitting smoke [82] and steam on occasion, sometimes dropping cinders or leaking water on to the road. Nothing was more aurally discreet than an electric delivery truck or van; nothing on wheels had less personality.

The sounds of cars were a means of identification even when the vehicle itself was still invisible. The most readily recognisable (and easiest, by far, to imitate) was the 'Bullnose' Morris, which made friendly, 'Pooh'-like hums in first and second speeds. Another easy one was the Austin Twelve, which had a singular selection of rather high-pitched hums in its indirects: but it took a real expert to determine the difference between a 14/40 Humber and a 3-litre Bentley if gear noise alone was used for identification. There was no problem if exhaust and engine noise were taken into account, the rich, deep-noted rumble of an accelerating Bentley being quite

80 *The crews of commercial vehicles inadequately equipped against the elements were not alone in the early 'twenties; on occasion the paid drivers of luxurious motor-cars found themselves in a vulnerable situation. This circa* 1923 *Daimler open-drive limousine, by Barker & Co, has bold lines and is above the average in appearance. Features of note are the 'opera' lamps at the front of the body, the unusually low-set speaking-tube, the blinds on the rear-door windows, the very positive curve of the back of the body and the wooden slats protecting the petrol tank, for long a Daimler identification (above).*

81 *The useful life of this Halley bus, and others like it, was greatly extended after the introduction of pneumatic tyres, its mechanism and bodywork being subjected to less transmitted shock and vibration from bad road surfaces. With so much window area the interior would be very light and airy (below, left).*

82 *A wisp of smoke, a puff of steam and the distinctive, well-rounded front of a Super Sentinel (in this case a well-restored* 1924 *model) bring back nostalgic memories to those who recall vehicles of this type in daily use well into the 'thirties (below, right).*

83 *Old Rolls-Royces, like old soldiers, never die. Even after countless thousand miles of daily use in Edinburgh, engine, chassis and body of this 1923 three-speed Rolls-Royce Twenty were still in first-class order when it was sold back into private ownership in the late 'fifties. The lightness of appearance of its body suggests that it may be of Park Ward manufacture.*

unmistakable. Most cars of lesser performance went about in a state of mechanical calm; some sang sweetly to themselves in their indirects and moved peacefully in top. The real motoring mutes of those years were the costliest cars—the big Rolls-Royces and Daimlers, Lanchesters, Sunbeams and the 40/50 Napier, for example. In its class, however, the 'Bullnose' Morris must be singled out for the quietness of its running at top speed. Rolls-Royces slid through cities with catlike quiet. The largest model, the 40/50, spent so little running time in its indirects, even in densest traffic, that few were ever aware of their mellow, musical boom—a characteristic inherited from Manchester-built forebears. The smaller Twenty, which had been introduced in 1922, had an extremely quiet engine and transmission and, as time would prove, the imperviousness to wear of the larger model.

Well into the 1950s, when numerous Twenties made between 1923 [83] and 1929 were still in everyday use as taxis in Edinburgh, it was an easy matter to judge at first-hand how splendidly the designers, gear-cutters, fitters and test staff at Derby had carried out their respective tasks. The Scottish capital has its share of hills, and at that period a large number of the thoroughfares in its centre were still roughly paved with granite setts. Tourists have always added greatly to the work-load imposed upon its taxi-cabs, and those elderly Rolls-Royces seldom rested. Even after many years of use in such conditions, (apart from time previously spent in private ownership), engines and transmission retained their original quietness in almost every case.

Maintained quietness was a feature, too, of certain bodies and that was an especially notable reminder of the high quality of design and construction provided in the first place by their makers. The unknown—and certainly unsung—craftsmen of H. J. Mulliner and Park Ward, the coachbuilders most concerned, carried out their work so well that even long years of door-opening and closing, scores of times each day, and countless thousands of miles over uneven granite setts had not disturbed the structural integrity of the bodywork. Standards so high were difficult and costly to attain, of course, and not every manufacturer sought to match them. As the decade progressed, however, there was a general move towards the production of quieter cars and this extended well down the price scale. Aural differences between the most costly and the cheaper kinds of car were very great indeed. It was not until the 'thirties that the somewhat narrower gap between medium-price machines and the most expensive was reduced significantly. By then much had been learned about the reduction of mechanical sounds at source, and about effective but inexpensive means of isolating them from vehicle interiors.

In 1919 however, the year in which the Institute of British Carriage Manufacturers became the Institute of British Carriage and Automobile Manufacturers (IBCAM), more serious matters faced the industries that it represented. Motor-car making for private users had almost stopped by 1916. The number of concerns allowed to satisfy Service requirements for staff cars, ambulances and so on, was small. The rest—the majority—had been directed to other forms of work, with a consequent rearrangement and total disruption of their normal production practices. Once the war was over the process had to be reversed, a time-consuming and costly matter. Raw materials, finished components, new machine-tools to replace those worn out in armaments manufacture, and skilled men had to be found. There was the additional complication of competition from newcomers to the industry. To make matters worse, there was a good deal of labour unrest for a time which, along with all the other difficulties, made it quite impossible to maintain regularity of output. Frustration was widespread among manufacturers, sales staff and distributors who had to deal with car-hungry customers clamouring for early delivery, without seeming to understand why this should be so hard to achieve [84]. There was also the matter of inflation, which made prices rise to such an extent that many potential buyers, whose budget was already a marginal one, had to reconsider their position.

84 *Bodybuilding was a generally agreeable occupation, especially when it was carried out in surroundings as clean and well-lit as these. The place is the hood-making department of Wolseley's Birmingham plant, in the early 'twenties, and the scene is typical with plenty of space for marking-out, cutting, sewing and fitting; all of it by skilled hands and eyes.*

In general, 1919 prices were twice as high as those of 1914, sometimes more. In that year the Rover Twelve five-seat touring car, comprehensively equipped (which was unusual at that time), was excellent value at £350. Five years later its successor, facelifted but otherwise substantially unchanged, cost £750 complete. The pre-war price of a Rolls-Royce chassis had been £985 and a complete closed car cost around £1,500. By 1919 the chassis price had risen to £1,450 and within twelve months it had risen again, to £1,865. By then a complete closed car cost about £3,000. The chassis price reached £2,100 eventually and did not fall until 1922; thereafter it remained constant, at £1,850, for the 12 ft-wheelbase model of the 40/50 until 1932. In 1920 £3,000 would buy one Rolls-Royce, twelve Model 'T' Fords or five luxury-model Morris Oxfords. In 1939 it would still buy one Rolls-Royce limousine, twenty-four Austin Seven saloons, fifteen Morris Twelves or eleven 8-cylinder Fords. In post-first war Britain, however, it was hard enough to acquire almost any kind of newly-made motor-car, and for a year or so second-hand values took account of current shortages. Very high prices were paid for ex-Services vehicles that had often seen an enormous amount of use, a much-criticised organisation called the Government Property Disposal Board being responsible for their sale. Immediately after major wars, it seems, many people have a great deal of money to dispose of. £3,000 and more was being paid for Rolls-Royces at auction in 1919, and even as late as November 1920 a 1914 40/50 was on offer at £2,500.

By 1920 to 1921, the post-war boom was over. A serious economic slump prevailed, and not only in Britain; most other countries were just as badly affected. Among its causes were widespread labour unrest and uncertainty, continuing problems of supply as far as raw materials and finished components were concerned, unwise investment and persistent inflation. The makers and distributors of motor vehicles were as much affected as everyone else, yet their present and long-term prospects were strong.

Before the war, ownership and use of motor-cars had been the privilege of the prosperous, on the whole, and the paid driver held a rather special place in the social hierarchy. The greatly expanded use of the motor vehicle during the war familiarised enormous numbers of people with its capabilities, at first hand, in Britain and elsewhere in Europe. As a result, attitudes towards it changed quite radically: it became less exclusive and more accessible in general terms than hitherto. In this respect European thinking came into line with transatlantic attitudes: In the United States the automobile had been taken for granted from earliest years as an economic and social necessity, seldom used to reflect its owner's prosperity or progressive thinking.

It was this new fact of life which encouraged the more adventurous and optimistic elements to plan for increased production from 1919 onwards and, while not all their schemes succeeded, for one reason or another, the actual achievements of the British industry were impressive. This applied to all areas of activity; to the making of cars, lorries, buses, bodywork [84 and 85], tyres and other components. But it is scarcely possible to overstate the number and the nature of the practical problems to be overcome. Even before the Armistice, established manufacturers had difficult decisions to make. It was reasonable to assume a huge demand for their products once

85 *Body assembly for a batch of Karrier lorries at the Martin Walter coachworks in the mid-twenties. Simple line-production techniques were used to speed up construction when the number of identical bodies going through the shops justified it.*

peacetime conditions returned, and to anticipate that in a situation of such shortages the reissue, as it were, of pre-war designs would be commercially acceptable for a time. By those with machinery not entirely worn out by munitions-making, the temptation to begin again by building 1914 motor-cars as soon as possible was not always resisted. Others thought differently and, by driving themselves and their staffs to superhuman efforts to design, test and manufacture vehicles, they were able to secure the attention of the press and the interest of potential purchasers.

At one end of the scale Rolls-Royce brought back their 40/50 chassis unchanged, except in details and in price, and with a greatly increased reputation as a result of its maker's wartime achievements on land and in the air. At the other, the Ford 'T', manufacture of which had continued in its native America throughout the war, reappeared on the British market with electric lighting and starting and a black-shell radiator (the latter a 1916 economy measure). Otherwise it, too, was unaltered; the same quirky, unconventional, sometimes exasperating, endlessly useful vehicle it had always been.

The Morris Oxford and Cowley—the latter a new model introduced in 1915, and made without a break thereafter until the supply of American-made components dried up—now shared the same engine and chassis, differing only in terms of coachwork, equipment and price. The post-war Cowley was of all-British construction. William Morris, who had 300 cars either completed or at an advanced stage of construction by the Armistice, had none left for sale by May 1919, and was thinking then in terms of an annual production of 5,000. It was to be some time before he achieved that aim. Rover hedged its bets by bringing back its splendid Twelve in slightly modernised form and by presenting an entirely new economy-type Eight, the design of which had been bought from J. Y. Sangster, a member of a family with long-standing motor-cycle interests. The Twelve was still aimed at the professional man market in which it had been so successful before the war and in

86 *The second famous Rover Eight was every bit as simple in design and construction as its Edwardian forebear, designed with great ingenuity to be easily and inexpensively manufactured, and to be reliable and economical in use. Its little body was not unattractive and it was, of course, turned out in quantity, quickly and cheaply.*

which its balance of qualities was so greatly appreciated. By comparison, the Eight was an ugly, noisy duckling intended for a different kind of user altogether. It was no more attractive in appearance than most utility cars of its day, and while its construction was in no way skimped, the little vehicle was clearly built to a precisely calculated price. When it was announced in 1919 this was £230: by 1920, however, it was up to £300. In March 1921 it fell to £220; in 1923 it dropped again to £180, then to £139, and in 1924 both catalogued models, the standard two-seater and the two-adult two-child 'Chummy', were down to £145. By then there was competition from the £165 Austin Seven, a miniature car of superior refinement of running. Most unusually the price of the Eight went up to £160 in 1925, the year in which it was being replaced by a new Nine. In its time the Eight provided many thousands of newcomers to motoring with mobility, reliability (once early engine problems had been resolved), and economy. Forty-five mpg was possible with care and under ideal conditions the 2-cylinder, horizontally-opposed, air-cooled engine would drag it up to 45 mph, which was more than twice the legal limit, and more than enough for most of its owners. It was a simple machine that deserved its major share of the economy car market. The Rover Eight [86] was a good bridge between the motor-cycle and the motor-car proper, and despite its relatively low first cost it was remarkably long-lived, sizeable numbers remaining in use until the early 'thirties.

Nothing quite like it followed until the introduction to Britain of the equally awkward-looking 2cv Citroën in the early 'fifties, the more shapely Volkswagen at about the same time and the highly attractive Fiat 500 in 1956. Although these cars were produced at different times and differed greatly in technology they shared many common characteristics. Each was designed with great attention to ease, to speed

87 *Apart from serving owners well in its day, the Trojan has been popular with post-Second-War trials enthusiasts, its ability to crawl up apparently impossible tracks being rivalled only by 2cv Citroëns and specialised cross-country vehicles. The punt-like base is still being used, much modified, by more than one present-day car manufacturer.*

and to economy of production: the Rover by the thousand; the French, German and Italian cars by the million. Each was well made, of excellent materials, for long-term reliability. Each had an air-cooled engine and like the first truly universal car, the Model 'T', each had strong elements of the unconventional in its design. This had no adverse effect on sales. As long as the ends sought are successfully realised in practice, it seems, deviations from mechanical orthodoxy can be overlooked, by customers and cost accountants alike.

The Ford 'T' actively defied many of the automobile engineering conventions of its time; what is more, it had a manner of moving and a look that were unique, to say the least. Yet, because of its positive qualities, more than fifteen million customers signified approval during its nineteen-year production run.

Between the wars no vehicle more original in design or, for that matter, more deliberately different in running and appearance than the Trojan [87] was successfully marketed in this country. Leslie Hounsfield, its designer, was a man of exceptional mechanical inventiveness and ingenuity whose attempts to create a car of outstanding simplicity and economy of operation had begun in the early 1900s and had reached the prototype stage by 1910. Although the Trojan company was formed in 1914 and, according to the late Anthony Bird,* put together five pre-production cars by 1920, actual manufacture did not begin until 1922, when it was in the hands of Leyland Motors Ltd. The normal products of that large Lancashire concern had raised it to an exalted place in the goods vehicle and motor omnibus hierarchy, through high standards of construction, exceptional reliability and

* Anthony Bird, *The Trojan Utility Car*, Profile Publications Ltd, 1967.

88 *The van version of the Trojan lacked elegance but its interior was roomy and its flat sides were excellent for display lettering.*

longevity. But their design was along conventional lines on the whole. This could scarcely be said of the ugly little vehicles made under licence from Trojan Ltd.

It was said that the engine of the Trojan had only seven moving parts, but because it lay beneath the passenger's seat, on the nearside of the vehicle, where it remained unseen but not unheard, and seldom seemed to need attention, few can have bothered to verify that fact. The engine was a 4-cylinder two-stroke of the so-called 'duplex' type, its paired cylinders arranged side-by-side so that they made a square-four monobloc. This highly unorthodox unit was connected to a set of three epicyclic gear-drums, with attendant brake-bands, which provided two forward speeds and reverse, and rested on the punt-like pressed-steel frame on its offside, below the driver. The final-drive was taken through double-reduction spur gears and single chain to a sprocket on the live rear axle. To ride even a short way in one of these singular machines is enough to persuade the most reactionary enthusiast for past achievements that the explosion engine and its appurtenances have made some progress since the 'twenties. The all-engulfing sounds and mechanical palpitatations rising from below on moving off are startling, to say the least, but they are reduced appreciably once the higher speed is engaged. The experience is reminiscent of take-off in a light aeroplane of the period, but without its exhilarating speed and life. On the move, the motion of the car on its long and supple cantilever springs, even with solid tyres on its steel disc-wheels, is not unlike that of a bluff-bowed boat riding a gentle swell. Did the ingenious designers of the 2cv Citroën ever take a close look at a Trojan, one wonders?

In terms of simplicity, operational economy and longevity, the Trojan amply satisfied its designer's aims. It was wise of Leyland Motors to undertake its manufacture when they did: for one thing it did not involve the company in major

XII *The two-seater was one of the most appealing of all cars built in the vintage period, between 1920 and 1930, and this Rolls-Royce Twenty, bodied by Claude Hamilton, of Aberdeen, and still in regular use, proves the point.*

XIII *The simpler shapes of cars and clothes in the 1920s gave relief to eyes weary, perhaps, of the fulsome appearance of so many in the pre-war years. This 1924 Audi open-drive coupé—an elegant machine—is socially correct in every respect.*

XIV *The average American automobile in 1925 had a confident look; its design was strong, its shapes well defined. A tall radiator, boldly curved mudguards, prominent lamps and bumpers, along with a sun-visor above the windscreen, were unmistakably transatlantic.*

89 *Many commercial operators made every use of their vehicles for publicity purposes. What is now the Esso company operated this eye-catching Model 'T' Ford van in the 'twenties.*

90 *To turn a car into a commercial vehicle was not difficult when there was a substantial chassis as foundation. This 1926 Morris light-van was still in regular use thirty years after it was built.*

capital expenditure at a time of general financial difficulties. With their high reputation it was easier to present an unconventional vehicle as a practical proposition, and with their experience and wide connections in the transport world they were very well placed to take a worthwhile share of the expanding market for light cars and small delivery vehicles. In van form, particularly [88], the Trojan was a great success, not only with the owners of one-man businesses but with large-scale fleet operators as well. Hundreds of red Trojan vans thrummed the length and breadth of Britain during the 'thirties laden with Brooke Bond tea. From its first appearance, however, the Trojan faced powerful competition. To begin with, this came from the Ford 'T' van [89] and one-ton truck, but from 1924 onwards William Morris joined in with his 'Bullnose'-based light-van [90] and ton-truck and, at a later date, with a 5 cwt van version of his lively little Minor. It was entirely logical that

91 *The sight of lorries riding pick-a-back is commonplace nowadays but it was relatively rare in 1922. The larger Renault has its driver sitting on top of the engine casing which, in turn, is above the front axle.*

passenger car chassis should be adapted as necessary for commercial use; the idea had been under development in the United States for quite some time but took much longer to become accepted in Europe.

It was taken seriously enough by Renault soon after the war as part of an ambitious, well-planned policy of making and marketing an enormously varied range of road vehicles [91 and XI], from the very small to the very large. No manufacturer in Britain sought to do the same. In Europe as a whole the only concern to rival Renault in this respect was Fiat, which had been established in Turin in 1899 to manufacture motor-cars. With the energetic, far-seeing and determined Giovanni Agnelli in charge, however, it was not long before the range of its interests was enlarged, and by 1919 Fiat was already internationally known for the design and building of cars, lorries, buses and bodywork, aeroplanes and aero-engines, ships, motor-boats and marine engines, as well as for the production of raw and processed materials. Although the Fiat organisation was not then quite as large as the Ford Motor Company upon which it was obviously modelled, it was already a major force in the Italian economy.

In Britain road-vehicle manufacture was not the main concern of any really large industrial group; for such companies as Sir Armstrong Whitworth, Beardmore, BSA or Vickers, (all involved in armaments-making, incidentally, and closely linked in other ways as well), it had been—or was—a sideline only. By their drive, courage and personal qualities, and because of the high quality of their products Sir Herbert Austin and Mr William Morris were able to establish themselves and their companies as leaders of the British motor industry by the mid-'twenties and it was not until the 'thirties that their positions were challenged in any serious way. Ford, Standard and

Vauxhall were then taking much larger shares of the family-car market, and the somewhat smaller Rootes Group was having considerable success in the popular 10 hp class with its well-rounded, well-liked Minx.

In 1919 Austin, Morris and everyone else involved in the industry were far too busy to bother overmuch about what might lie far ahead. Certainly all were hopeful as far as future prospects were concerned, and more than one of the older companies, along with certain of the newcomers, had publicised ambitious schemes for expansion. It was scarcely a secret that the answer to the problem of making cars that people could afford to buy was to make large numbers in the most economic way possible. There was always the example of America to follow, but there was a gigantic difference of scale to take into account. In 1913, for example, 1,200,000 vehicles were registered in the States. In Britain, in 1914, the total number of road vehicles in use was a little below 264,000; by 1919 it had dropped to about 216,000. By 1926, however, it had passed the million, and in that year 153,000 cars and 44,500 lorries and buses were built in British factories. As a comparison, the Italian total for the same year was 63,800 vehicles, of which 34,200 were exported. In 1909, long before mass-production proper, the Ford Motor Company was turning out one hundred cars a day. Between July 1914 and July of the following year it built and delivered 326,933 vehicles. But the record-breaking year for Model 'T' production was 1923, when 2,019,000 cars, trucks and light delivery vans were completed. That amounted to approximately half the total production of the entire North American motor industry for the year, and was an extraordinary achievement by any standards. By comparison, European accomplishments and forward prospects seemed insignificant. They had a great deal of catching-up to do.

Here reference must be made to the continuing importance, on the American automobile scene, of the top-grade car, often carrying specially built coachwork from one or other of the leading concerns of the day, during this decade and the next. Although the number turned out was great by European standards it was minute when compared with the quantities produced daily by Chevrolet, Dodge or Ford. Nevertheless the significance of these fine cars, qualitatively and historically speaking, was considerable, and the reputations of such makes as Cadillac, Duesenberg, Lincoln, Packard, Peerless, Pierce-Arrow and Stutz, for example, were justly earned. So, too, were those of custom coachbuilders of the standing of Brewster, Derham, Fisher, Fleetwood, Hibbard and Darrin, Le Baron, Murphy and Willoughby. From a constructional point of view they matched Europe's best, and during this period a distinctively American style evolved, which shared certain characteristics with bodies stamped out by the hundreds of thousands in the great press-shops of the American industry. Some of these will be noted in plates 91a and 91b.

Although doubts were expressed from time to time about the capacity of the British home market to absorb the increased output of an enlarged industry once conditions were back to normal, the general feeling was optimistic. Major R. Vernon C. Brook, the level-headed writer of an article that appeared in *The Autocar* of 6 September 1919, estimated that the shortage of new British-built cars at that time amounted to 250,000 and that there was a further shortage of 80,000 imported vehicles (earlier in 1919 it had been calculated that the shortage of cars in the United States was

91a *This 1924 Lincoln Model L has a custom-built Limousine body by Fleetwood. The very high quality of its finish is apparent, even in a photograph, and it is obvious that great trouble has been taken over its appearance design. The bumpers, the drum-shaped headlamps and prominent tail-lamp, the sun visor, the steel-disc wheels (optional) and the clean lines—all are uniquely American.*

91b *Even in 1925 the look of this Lincoln was somewhat archaic. Its body was designed by Wellington Everett Miller, built by Murphy, of Passadena, and commissioned by Mrs Rudolph Spreckels. Such details as the lamps, radiator, bonnet, wheels and controls are up-to-date; they are visually acceptable and so, too, is the passenger accommodation. It is their attempted union that is not altogether satisfactory.*

700,000). Some of the thirty-five home manufacturers whose annual output in 1913 had exceeded 500 units each had announced their intention of producing upwards of 6,000 cars a year, and ten, at least, of the new concerns were talking in terms of 3,000 to 6,000 annually. Nevertheless, Brook estimated that to satisfy current demand alone would take two years. As it happened the problem solved itself, more or less: production did pick up in 1920 and much of the backlog had been dealt with by the time the post-war recession began to take effect towards the end of that year. When it was over prices were still going down: production difficulties, it seemed, were over; orders were tumbling in, and while demand for new cars was in no way as clamant as it had been, it was increasing steadily. It was now quite clear that the industry as a whole was set for a spell of expansion and prosperity. It was also obvious that it was becoming a more and more important element in the British economy.

All the same, the slump hit some car makers very hard. The weakest dropped out of a struggle for which more often than not, they were inadequately equipped; their organisation, management, financing, selling methods or the product itself being deficient in one way or another. One or two were simply unlucky. Some of the older concerns and more of the newer ones gave up. For example, eight of the former and fourteen of the latter closed down in 1922, and another six and sixteen respectively, in 1923. In 1924 the figures were seven and twenty, and in 1926 six and twelve. After that the annual rate of failures fell. It was to be expected that cars of inferior design and construction would disappear. What was not anticipated, perhaps, was the short life of some of the really good ones.

The many makes that survived usually deserved to do so. The rag, tag and bob-tail that appeared after the war had too much of the makeshift about them; frail bridges between the motor-cycle and the motor-car that lacked the life of the former and the comfort of the latter in almost every case. Two of the few exceptions were the Rover Eight, which was made by a most reputable company and backed by a sound service organisation, and the more dashing GN [92], a cyclecar of pre-war origin with character, real individuality of design and, in its most potent forms, exciting performance. It outlasted pre- and post-war rivals, and although its day was done by 1924 certain elements of its design lived on in the later Frazer Nash and HRG sports-cars.

92 *One cyclecar that survived the war and resisted competition from the new light-cars for quite a time was the GN; in its final form it was a nicely-balanced little machine and a handsome one.*

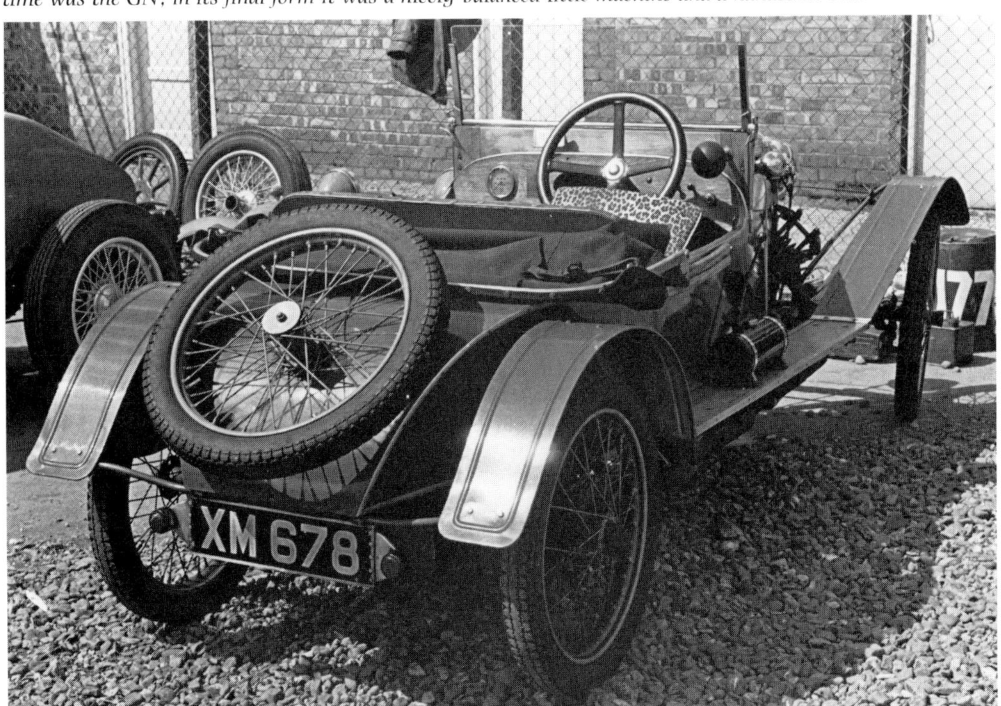

Others promised well on paper or at the prototype stage but failed to reach production for one reason or another. Unlike so many of its contemporaries the Cosmos, or CAR, had a sound technical background; A. H. R. Fedden was among those involved in its design and production was in the hands of the Cosmos Engineering Co, of Bristol. This concern (formerly Straker & Co, the makers of the Straker-Squire car*) specialised in the design and manufacture of aero-engines and it was not surprising, therefore, that the engine of the Cosmos—'a utility car for the man in the street', as contemporary publicity material described it—was air-cooled also. It was of the stationary radial type; in-unit with a three-speed and reverse gearbox and with its three cylinders arranged in inverted Y formation. Its official rating was 11.9 hp, its radiator was a dummy, its rear suspension was by coil-springs and bell-cranks and its wheels were ugly Goodyear 'wavy discs'. Production of 200 cars a year was expected by September 1920. *The Autocar* reported favourably on a journey in a pre-production model, and a light-blue car, with pigskin upholstery and polished mahogany cabinents at either side of the rear seat, was exhibited at the 1919 Motor Show. It is not known if its body was made of papier-maché, as intended by Roy Fedden—always an inventive individual. The make did not reappear.

Another contemporary with an air-cooled stationary radial was the Enfield-Alldays, which had a 5-cylinder, 1,450 cc engine at the time of its introduction. Because of supply problems with the cylinders, it was said, this intriguing unit had to be abandoned and a more conventional 4-cylinder engine took its place when production began. The little Manchester-made Duplex also started with one kind of engine (a most unusual 8-cylinder, of only 1,478 cc) and went into production with a commonplace four.

The 1922 North-Lucas was inspired to some degree by the highly original and interesting Rumpler, from Germany. It had a 15 hp, 5-cylinder, air-cooled static radial engine, its chassis and body were combined and it had separate suspension for each wheel. It was as unconventional in appearance as it was unorthodox mechanically, and it was this, as much as anything else, that prevented commercial success. One example survives. Some years later the larger, more ambitious Burney Streamline, which was also out of the ordinary in design and appearance ran into similar problems. It was the work of Sir Dennistoun Burney, the designer of the airship, R100.

Not that normality necessarily guaranteed survival either. One of the most interesting of the new post-war cars was the 40/50 Napier [93] that appeared in the autumn of 1919, a product of one of the oldest firms in the British industry, and a direct challenge to the long supremacy of the 40/50 Rolls-Royce. Its main feature was its beautifully-made light-alloy, 6-cylinder engine, an up-to-the-minute unit that owed much to Napier's wartime experience of aero-engine design and construction. But it was not of the high-efficiency type that might have been expected from a company with so much racing and record-breaking in its past—and by then there was no S. F. Edge on hand to explain this away in cunningly written

* A. H. R. Fedden was also responsible for the design of the post-war Straker-Squire.

letters to the motoring journals. He was otherwise engaged; about to beat loud drums for the new AC Six that was to be built in moderate quantities for the highly critical small car owner. The 6,177 cc engine of the Napier developed 82 bhp at 2,000 rpm (13.29 bhp per litre) and could not stand comparison in that respect with its contemporary, the 37.2 Hispano-Suiza. The engine of this magnificent new car was, in effect, one half of a projected V-12 aero-engine and produced 135 bhp at 3,000 rpm, or 20.45 bhp per litre. The closest English contender in the same class was Parry Thomas's Leyland Eight, the 7,266 cc engine of which gave off 145 bhp–19.96 bhp per litre.

Surprisingly, in view of its reputation in later years for sweetness and sloth, the Rolls-Royce Twenty [XII] shows up well, comparatively speaking. If factory figures are to be believed its engine developed 52 bhp at 3,000 rpm, or 16.61 bhp per litre. Corresponding figures for the Lanchester Forty, the other major rival to the 40/50 Rolls-Royce lead in the luxury car class, were 95 bhp and 15.38 bhp per litre, respectively.

Power as such was not the main aim of Lanchester, Napier or Rolls-Royce, of course; these makers were more concerned with mechanical good manners, with quietness and extreme flexibility of movement. As it happened, the Lanchester and the Rolls-Royce were quick as well as quiet. Although the Napier pulled a very high top gear its maximum was little more than 65 mph; perhaps it was as well that it was no higher, for chassis design and the braking system lagged somewhat behind the quality of its engine. Mr Ronald Barker, a Napier owner and user, an engineer and a responsible historian, has suggested that A. J. Rowledge, who was responsible for the overall design of the 40/50, may have been out of his depth when he came to deal with chassis and transmission arrangements and may have delegated work on those

93 *Few 40/50 Napier cars had real quality of appearance when complete and this was the result of poor chassis appearance in the first place. Their proportions were unsatisfactory and this fact was reflected in the awkward look of most finished cars.*

94 *George Lanchester's Forty hp model was an excellent car but its appearance was clumsy and while it had distinction it lacked the sort of style that sold so many Rolls-Royces and Hispano-Suizas. Lanchester's own bodies were of first-class design and construction; all they lacked was grace of line and proportion.*

vital elements to a subordinate, or simply failed to understand the importance of bringing them up to date. Certainly he was not involved in chassis design once he moved to Rolls-Royce.

Mechanically speaking, therefore, the new Napier lacked the balance of qualities that distinguished the Rolls-Royce from all its rivals; furthermore it did not have the sheer physical presence of the Derby car: its radiator was more noticeable than notable, and the proportions of the car overall were never altogether satisfactory, despite the best efforts of the Cunard Motor and Carriage Co, an associate company responsible for most Napier bodies. Its best-looking feature, its engine, was seldom seen.

The Lanchester Forty [94] had a much better blend of qualities. Its appearance was impressive. It had an excellent overhead-camshaft engine, its transmission incorporated an easily managed three-speed and reverse epicyclic gearbox and final-drive was by its maker's noted worm and wheel. It had a braking system as effective as anything offered by its English peers. In this last respect, however, no British product could compete with the best then available from France. Perhaps four-wheel braking became so popular there because its roads and its lack of speed limits made fast motoring so easy and so enjoyable. French roads had been built by Napoleon for the swift movement of armies; they were straight and they facilitated rapid movement. The fact that so many French road surfaces were poor—a direct consequence in northern areas, of wartime damage—was a mighty stimulus to improvements in chassis and suspension design. Since a lack of speed restrictions meant that high speeds could be maintained over long periods, special attention had to be given to the development of engines with the necessary mechanical stamina.

On the much busier roads of post-war Britain speedy cars could be at a disadvantage, and four-wheel braking, pioneered in this country many years before, then dropped because of prejudice generated, it has been said, by jealous rivals, had to be taken seriously again. In 1910 Arrol-Johnston and Crossley cars were offered with four-wheel braking based on Allen-Liversidge patents, and in 1911 Argyll introduced a diagonal system, designed by J. Rubury, that remained in production much longer. It was ironic that Britain now had to make a new start in this matter. Not that anyone seemed to be in much of a hurry to make the first move. The lead came, somewhat surprisingly, from Sir Herbert Austin, whose Seven, first seen in public in 1922, had a tiny brake-drum on each wheel. During 1923 Bentley and Sunbeam offered front-wheel braking, as did Palladium, which was far less costly, and in the following year they were joined by Daimler, Lanchester, Napier and Rolls-Royce, among others. It was very much a matter of the pace of progress catching up with the pace of traffic. By 1925 the fitting of four-wheel brakes to cars was widespread in Britain [95].

95 *William Morris (later Lord Nuffield) was well aware of the value of good publicity, and during the 'twenties in particular his company's press advertising maintained high standards. In this February 1925 announcement the virtues of Morris four-wheel braking are vividly proclaimed.*

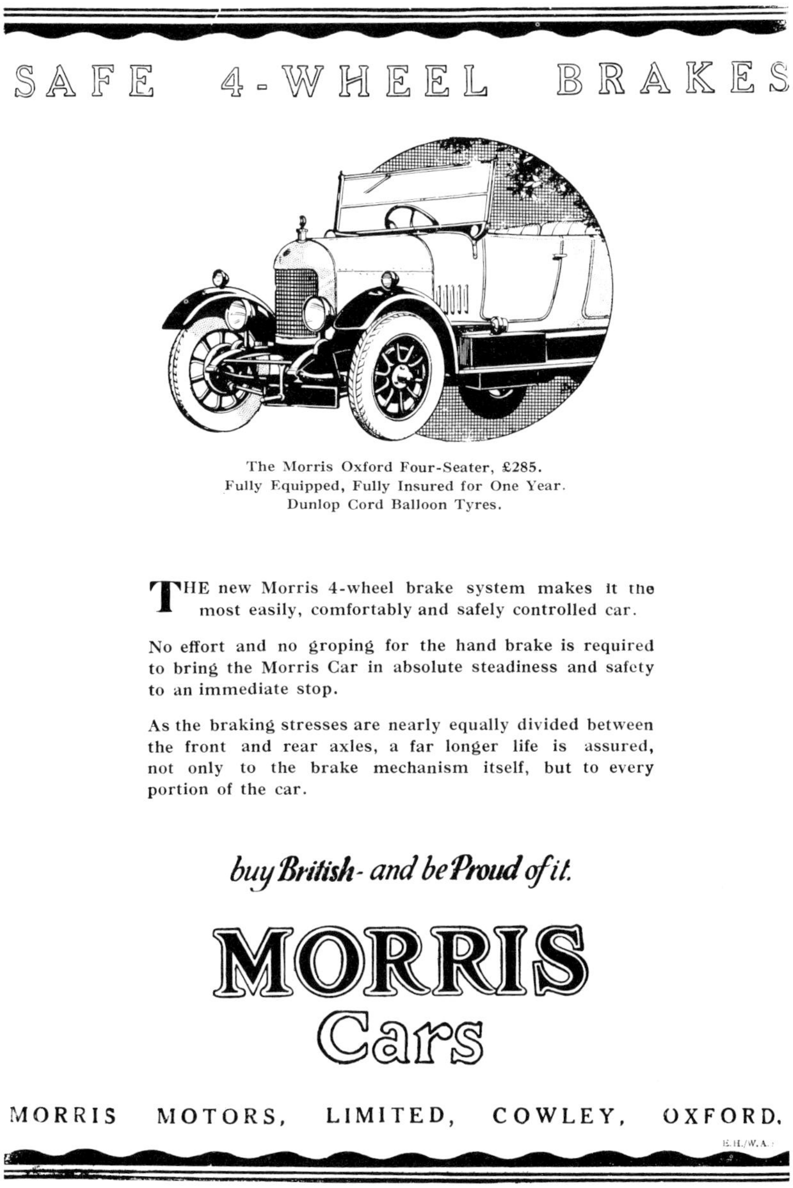

96 *The 1920 Leyland Eight was not a handsome car but it had an individual appearance, with its curled mudguards, its large conical hub covers, its strangely formed lamps based on the shape of Parry Thomas's wristwatch, its vertical radiator, bonnet louvres and windscreen, and its near-vertical steering wheel.*

At the same time more and more bus builders were improving the braking of their chassis. Steady improvements in engine efficiency were pushing up performance all the while. The change to pneumatic tyres encouraged drivers to make increasing use of the speed potential of their buses, a fact angrily noted in letters to the motoring journals. Traffic densities were rising rapidly. From 1925 onwards four-wheel brakes became standard equipment on an increasing number of makes of bus, often with vacuum servo assistance; at last the driver's task was being lightened appreciably. Braking on all road wheels followed more slowly on goods vehicles; indeed it was still an optional extra on some chassis at the beginning of the next decade.

One English car had vacuum servo braking well before its contemporaries, along with other features much ahead of their time. The Leyland Eight [96] created much interest and excitement at the 1920 Motor Show; technically speaking it was forward-looking and, while its appearance was in no way handsome, it had individuality. Its in-line 8-cylinder engine was superbly finished, with the external neatness of the best kind of engine from Europe, and it was the most powerful touring car unit available anywhere. It had a very high performance in standard form and, as its designer, Parry Thomas, demonstrated frequently at Brooklands, the mildest of tuning turned it into a most potent machine. With so much speed available it is rather surprising that Thomas confined braking to the rear wheels only.

The engine of the Eight had a one-piece detachable cylinder head. Its single overhead camshaft was driven by triple eccentrics, and there were two inclined tulip valves to each hemispherical combustion chamber, the opening and closing of each pair being controlled by a leaf spring. The crankshaft had six bearings. There was thermostatic control of the cooling system. Automatic chassis lubrication, fed by little plunger-pumps that were operated by movement of the rear springs, practically eliminated routine maintenance, normally a lengthy, messy business at that time. Torsion-bars formed part of the rear suspension, quarter-elliptic leaf springs the rest, and radius-rods, front and rear, positively located the axles. Unusually for a period when high-geared steering was the rule in Europe, even for the largest, weightiest kind of car, three turns of the wheel were required to swing the Eight from lock to lock. For all its size—wheelbase lengths of 10ft 6 in, 11 ft 9 in or 12 ft 6 in were offered—it was not too heavy, a chassis weight of 28 cwt being representative. The frame side-members and the main cross-member amidships were lightened by carefully dimensioned holes of varying diameters.

In retrospect it seems sad that a car so rich in ideas of a practical, forward-looking nature failed to fire the enthusiasm of more of the wealthy motorists of its day. Fewer than twenty chassis were built; not all were sold, and a courageous attempt to bring the English luxury car up to the moment had to be abandoned. Perhaps its specification was a little too advanced. Perhaps the name Leyland did not suit, in view of its commercial-vehicle associations. On the other hand, was the rather clumsy, ill-proportioned appearance of the Eight not to blame to some considerable extent?

The Lanchester Forty, another large luxury car that could be made to travel quickly with the minimum of tuning, fared rather better, about 400 examples being made and sold between 1919 and 1929, to be superseded by an excellent 30 hp Straight Eight. The Twenty-one was a smaller model, launched in 1923 to compete head-on with the new Rolls-Royce Twenty. Unlike its costlier rival, which did not acquire four-wheel brakes or a four-speed gearbox until 1925, George Lanchester's 'pup', as he called it, had these important features from the start, and because of a more vigorous engine it was a little faster and a little brisker in acceleration. Yet, it was neither so quiet nor so handsome as the Rolls-Royce; its rather stolid appearance lost some sales, surely, even though its chassis price was always £50 below that of the Rolls. Statistics may or may not tell the whole story, but between 1923 and 1929, when about 700 of the smaller Lanchesters were built (the majority bodied by their maker with customary quality of technical design, workmanship and finish) almost 3,000 Twenties went from Derby to one or other of the scores of coachbuilders on their maker's approved list.

Some emphasis has been placed, thus far, on certain of the more unorthodox machines of the early 'twenties, largely because they were more interesting technically than many of their conventional contemporaries. The latter predominated, of course, once the post-war scramble for customers had resolved itself; by then the price advantage enjoyed by the revived, revised cyclecars of pre-war origin, or of the new and strident air-cooled cars had ceased to prevail and they had to give way to water-cooling and the 4-cylinder engine. Some of the excitement—or exasperation—of motoring was lost, no doubt, but on the other hand journeys started were journeys completed, more often than before.

97 *Lorries of this age and type were in everyday use until the early 'fifties, particularly in and around London. Often, as here, these vehicles were attractively painted and lettered, and they were usually very well maintained. The open-sided, 'C'-type cab was typical.*

Levels of comfort for the occupants of commercial vehicles [97] rose steadily during this decade, and the same was true of passenger cars and motor buses. Open bodied cars were better protected against the weather, the Standard company showing the way with an intelligently designed system of snugly fitting hood and side-screens that approached saloon-car levels of insulation from the elements [98]. Seating and suspension were also improved, the new balloon tyres helping greatly in this respect. Engines and transmissions were quieter; so, too, was bodywork.

This last was of ever increasing importance as speeds rose. A corresponding improvement in road surfaces or in chassis rigidity did not, of course, necessarily follow, and it was more necessary than ever before to produce bodies sturdy enough to resist the multitude of strains and stresses received in fast driving over indifferent roads. The coachbuilders coped. As far as bodywork of composite construction was concerned, (wood framing, with steel or aluminium panelling), they had considerable experience to draw upon. When the new Angus-Sanderson [99] car was being designed for large-scale production, those responsible were working in asssociation with a North-East of England concern that had been building horse-drawn vehicles of high quality since the eighteenth century. Somewhat unusually in a craft notorious for conservatism, it was prepared to adopt new ideas and methods; to make extensive use of jigs and accurate patterns for the cutting of frame members and their assembly, followed by 'cladding' with preformed metal panels and trimming, all on a scale far beyond anything previously conceived. In that case a firm with long-standing traditions and an enormous store of practical knowledge readily adapted to new needs, and the pity is that such a generally sound project came to nothing in the end. The precedent had already been established by Henry Ford, whose techniques of large-scale production had been developed to an astonishingly high level of efficiency by 1914. In that year, with three parallel elevated-rail assembly lines in action, chassis were produced on an upper floor (where bought-in parts were assembled with others made on the premises at

98 *The patented hood and side curtains of the post-Armistice Standard tourers were both neat in appearance and efficient in use, light in weight and easy to manage.*

Highland park, and the upholstering, trimming and painting processes carried out), and lowered down chutes to be bolted to chassis waiting below, in a most carefully-synchronised operation. As far as painting was concerned, the most important coats were 'flowed-on' by gravity, as was the finishing varnish, and this method was followed in principle by Angus-Sanderson in 1919/20. Properly handled, it was quick and effective in mass-production.

99 *It is not surprising that the bodywork of the Angus-Sanderson was of good design and quality, for the manufacturers of this car had been coachbuilders since the eighteenth century. The general shape of the standard tourer on the 1921 14 hp chassis is notable for its subtlety and for the lack of that makeshift car characteristic of the early 'twenties. The body sides have pronounced 'tumble-home', to give occupants plenty of arm and shoulder room and the doors have concealed hinges within the standing pillars. It was intended for mass-production on an ambitious scale.*

100 *In 1924 the all-steel body lacked grace of shape and proportion but its basic shape was simple and straightforward, and it was improving. The Essex Six coach did much to popularise this type of construction and the balloon tyre.*

101 *Post-war Sunbeam cars maintained their maker's high reputation for quality of chassis and bodywork. The long, low body of this 1921 20/70 hp model is a barrel-sided tourer, with a deck between front and rear compartments. The large castellated tank filler cap was a distinctive Sunbeam feature.*

But the new all-steel bodies were another matter altogether, and until sufficient knowledge had been acquired of their characteristics (and of methods of dealing with the problems certain to arise in service) a wait-and-see policy had to be followed. Motorists were already aware of the fact that a touring car with its hood up was a good deal noisier than when the hood was down. In saloon bodies the situation was worse: occupants sat in what was essentially a mobile sound box [100] a large-scale amplifier of any noises or vibrations fed into it from mechanism, road surfaces and its passage through the atmosphere. In general terms, this fact had been obvious to body designers and builders for a very long time, and with the growing popularity of the closed car solutions had to be found.

Ever since increases in engine efficiency had made possible the mounting of closed coachwork on the more powerful sort of chassis, they had been bothered by the incidence of boom and drumming, roof and body panels (especially when these were metal) serving as effective receivers, converters and transmitters of sounds and vibrations generated by vehicle movement. In the case of the more powerful kind of 4-cylinder engine both noise and vibration could be heard and, in extreme cases, the latter was visible; at tick-over speeds, in particular, lamps, horns, mirrors and other similar accessories trembled violently. Apart from any strain on the nerves of the sensitive observer the effect of transmitted vibration of this kind could be cumulatively harmful to body structures. Some were almost shaken to pieces by it. By the time car production resumed in 1919 these problems were of less consequence, largely because of improvements in engine smoothness and quietness of running. As far as the more sporting type of car was concerned, however, it was still advisable to specify open bodywork [101]. Within a year of two matters changed for the better with the introduction to Britain of the Weymann system of closed-body construction from France.

The Weymann body had a lightweight, non-squeak wood frame, covered with tightly-stretched leathercloth of a special type and assembled in such a way that no two pieces could rub together (each wooden member was secured to, but separated from its neighbour by a metal plate), and the absence of metal panelling eliminated the booming still characteristic of so many closed bodies of traditional construction. Its inventor, M. Charles Weymann, claimed that bodywork built to his design would be lighter and quieter than that of conventional type and no more expensive to maintain or repair. From lengthy personal experience the writer can confirm the general truth of his statement. A Weymann body, fitted to a Rolls-Royce Twenty chassis, gave to the latter the 'life' intended by its designer in the first place but usually stifled by luxury-seeking owners. There were one or two practical drawbacks, however. In common with the pressed-steel system it imposed quite severe restraints on the designer; in particular, he had to make curves as flat as possible. The covering material used had a rather dull finish and was available in a restricted range of colours, and these factors had to be taken into account at the design stage as well. Such limitations did not prevent the creation of many Weymann-type bodies of outstanding quality, and in this respect the work of H. J. Mulliner [102] and J. Gurney Nutting & Co stands out for sustained levels of excellence.

102 *A flawless combination of engineering and coachbuilding skills; a 1929 Rolls-Royce Twenty fitted with a four-light Weymann, fabric-covered body by H. J. Mulliner Ltd. The contrast between the semi-matt finish of the special fabric (which is lightly textured), the high gloss of the coach-paint and the gleam of nickel-plating is subtly effective.*

Use of Weymann bodies was not confined to the more expensive car with bespoke bodywork. Its lightness and its freedom from boom made it attractive to the makers of cars of small horsepower and, as far as is known, the first of these to take out a licence from the sole British licencee, Rotax (Motor Accessories) Ltd, was the Rover Company, in 1923, shortly after the introduction of the Weymann system to the British motoring press, and a little before it was adopted by Talbot. Rover's bodybuilding department at the Parkside Works in Coventry soon mastered the niceties of the Weymann system and had a rather plain-looking saloon on the new Nine chassis in production by August 1924. For many years after Weymann-type closed bodies were available on all Rover chassis and sold well, their appearance gradually improved as a result of steady development.

By 1929 licences had been granted to some seventy-five coachbuilding concerns or departments in Britain alone, but in France, by then, the system had lost its former popularity in spite of efforts to improve its external finish in different ways. At the Paris Salon of that year the semi-Weymann was introduced—metal-panelled below the waist-line, leathercloth-covered above. It had a certain appeal and extended the acceptability of the system by a year or two more. Apart from the fashion aspect, which was an important element in its initial acceptance and an equally important one in its eventual rejection, the Weymann type of body made it possible for sports-car owners to enjoy saloon-car comfort in comparative quiet, without significant loss of liveliness. Its lightness helped here, just as it did with the much smaller, much less-powerful cars to which it was fitted. Lightness was a feature of the Gordon England system of construction that originated in this country [103 and 104].

Apart from anything else the system compelled the builders of bodies in the traditional manner to reconsider their methods, to find ways of reducing weight without losing strength of structure. In this respect the London firm of Park Ward was very successful; one six-light saloon built in 1927 with hollowed-out timbers in its framework and thin steel plates for joint reinforcement weighed no more than a Weymann body of similar type and size. It was a pity that the success of the system encouraged the production of cheap copies—fabric-covered bodies that were not always very well made and quickly deteriorated unless they were carefully maintained. Not all imitations were bad, of course [105]; indeed the style was developed most effectively during the second half of the 'twenties and fabric saloons,

103 *The completed frame of a Gordon England saloon body for an Austin Seven chassis,* circa *1927. Its components have been accurately machined to pattern in the wood mill and assembled by hand on this simple stand. At the next stage the doors will be hung, after which the framing will be covered, inside and out.*

104 *A Gordon England fabric-covered body on a 1929 Triumph Super Seven chassis. A feature of this patented system was lightness—the body alone weighed $2\frac{1}{4}$ cwt. The body shell was rigid but it was flexibly attached to the chassis at three points, while the floor and seats were secured independently to the frame.*

105 *The Chrysler was a popular American make in Britain and many chassis were imported to have English bodywork fitted. This 1929 '70' model has a Hoyal fabric-covered sports-saloon that has lasted for more than fifty years; its fabric may have been renewed on one or two occasions. Not unusually, even in 1929, wooden artillery-type wheels are fitted.*

103

104

105

106 *Sunbeam's own bodybuilding department specialised in the Weymann system and produced many first-rate examples during the 'twenties. This is a sports-saloon of exceptional quality and luxury, with wide doors and running boards, dummy hood-irons and a large, fitted trunk at the rear.*

107 *Even from a small photograph it is possible to gain a clear idea of the standards of workmanship maintained in the Sunbeam body-shops. The interior of this large Sunbeam close-coupled saloon is large enough to house occasional seats in the rear compartment, and there is a glass division separating it from the front. The woodwork is highly finished and perfectly fitted. Note the foot-rest ahead of the back seat; individual locks on the doors, the deep upholstery of the seats and the vertical pleating of the door trims.*

108 *A corner of the drawing office at Hooper & Co in the late 'twenties. It cannot have been easy to draw on paper placed vertically but there was plenty of light and space. Set-squares, extra-large French curves and other aids were plentiful.*

for example, featured in the catalogues of many manufacturers. Again their appeal was a rather specialised one. At its best, handled by top-grade makers of their own bodies like Rover or Sunbeam [106 and 107], or by outside concerns of the standing of Barker, Freestone & Webb, H. J. Mulliner or Vanden Plas, the type had great subtlety of surface texture, but the finish was less immediately appealing than the high gloss of the cellulose paints that began to replace traditional paints and varnishes from 1926 onwards.

This is not true of the best kinds of coachwork, however. Tradition continued to be followed as a matter of course, and rightly so; there was no substitute for ancient methods and well-tried materials in the hands of the master-craftsmen employed by the leading firms; no superior alternative to their skills in the shaping of body panels, surface-finishing, foundation preparation and patient applications of coach-paint, layer upon layer, with laborious but fastidious rubbing-down in between, before the final varnishings [108]. Perhaps the giant press-tools could stamp out steel panels as

107

108

109 *To photograph what is almost certainly a polished aluminium body is an extremely difficult task; in the circumstances the unknown Barker photographer did well. Because the name Patiala has been associated with the 40/50 hp Rolls-Royce illustrated, it is assumed that the car was built for an Indian ruler, which would explain the unusual finish. It has a deck between front and rear seats and wings of a most attractive type.*

110 *A better, clearer idea of surface finishes is to be gained from this photograph of a Hooper-bodied 40/50 hp Rolls-Royce,* circa *1923. Here the flawless painted surfaces contrast with the rich gleaming of nickel-plating and the colour and texture of the sham-cane covering of door and quarter panels. The body might be described as an all-weather, or convertible.*

smoothly surfaced as the hand-formed aluminium panelled of a bespoke body [109], but no mechanised process then known could match the flawless finish, the depth and richness of colour, the sheer quality of the best coach-painting [110].

Standards as high, as zealously maintained, applied to all other aspects of bodybuilding at that exalted level. Wood was widely used, not simply for the unseen bodyframe [111] but for functional and decorative details within, as a most important element in interior design; for instrument panels, for window and windscreen surrounds, for division and occasional-seat housings, for drinks cabinets

111 *The process of designing and making wooden bodyframes is difficult and complex; nowadays it is also extremely expensive. Accuracy is essential at every stage of an operation which might employ a gang of skilled bodybuilders for the best part of a week. This is a Sunbeam saloon body almost ready for panelling.*

112 *The formal town car; a circa 1926 Rolls-Royce Phantom I, with Hooper coupé de ville body. It is not at all a showy car, but it has character. The large leather hood is expertly cut and fitted. Keeping it polished and supple to the required degree, must have been one of the chauffeur's more difficult regular duties.*

and so on. The kinds of wood chosen depended, of course, on the taste of customers and varied widely, in consequence; figured walnut was a favourite but extremely rare and exotic timbers were specified at times, and high-gloss finishes were generally preferred. For obvious reasons, the interior woodwork of open bodywork was treated to make it weatherproof. Likewise leather was universally chosen for the upholstery of open bodies and it was also used extensively in closed cars [112 and 113], except those of the more costly kind, where it gave place to high-quality cloths (Bedford cord was popular, as was West of England cloth), or rarer kinds of material for rear

113 *Two trimmers at work at Hooper & Co. They are padding the hood of a large Packard before fitting the head leather itself; a task requiring enormous skill and patience but the simplest of tools.*

compartments. The luxury vehicle was still usually driven by an employee and not by the owner and the choice of black leather for the driver's seat made practical sense. If he returned to his place with clothing soiled from roadside repair work the marks transferred temporarily would not show and spoil its appearance.

The paid driver continued to play an important part in upper-class motoring. He it was who drove the rich far and fast on continental roads once the war was over, and he was a contributory cause to the problems faced by chassis and body manufacturers alike. H. I. F. Evernden, who served Rolls-Royce throughout the 'twenties as an exceptionally able and well-informed technical ambassador to their numerous accredited coachbuilders, has told of the difficulties then encountered.

114 *For many connoisseurs, the large open touring car is still the most cherished, the most longed-for. Here is a splendid specimen; a 1924 37.2 hp Hispano-Suiza fitted with a most handsome and comfortable body by Bligh Bros, of Canterbury. Features include the engine-turned fascia panel, the handsome instruments at front and rear, the V-screens and the broad deck between front and rear, the latter serving a double function as shelter and reinforcement for the body.*

115 *A barrel-sided tourer, by Park Ward, on a 1923 40/50 hp Rolls-Royce chassis. There is a strengthening 'bridge' between the seats as in 114 and again, the finish is superb.*

These were not of a mechanical nature, needless to say, but all to do with bodywork that simply shook itself apart in extreme examples, or gradually broke up under the stress of high-speed running for hours over the indifferent or poor road surfaces of favoured routes through France—to the Riviera, for example. Problems of this nature were not exclusive to Rolls-Royce; it just happens that their one-time body liaison engineer took the trouble to note them down.

Like the wheel-tramp difficulties that all makers of large and speedy motor-cars encountered once they began to fit low-pressure balloon tyres, these troubles were not talked about publicly at the time. Those intimately involved, however, could all tell tales of popping panel-pins, doors flying open (or off!) at speed, and body frames that came apart at their joints. To some extent the bodybuilders were to blame. They had been slow to respond to pleas from chassis manufacturers for a closer collaboration at the early stages of a new design; for a move away from the long-held opinion that chassis and body were separate parts, created quite independently of one another and without reference to special needs or practical problems from either side. Generally speaking, open bodies were less troublesome in service [114 and 115]. But the designers and makers of car chassis must also take their share of responsibility. Sound structures depend upon sound foundations, and these were not always provided for the bodybuilder.

At the beginning of the century Dr Lanchester had shown that frame stiffness was of prime importance, not only to secure as sound a base as possible for bodywork but to serve as a vital element in suspension design. In general, however, Lanchester's entirely logical ideas were ignored, though Rolls-Royce made an experimental car around 1911 that had unusually deep chassis side-members, and Parry Thomas took great care to secure maximum possible frame rigidity for his Eight when he was working on its design with Reid Railton, towards the end of the war. It was not until

116 *The Lancia Lambda had distinction of appearance but gave no external indication of its integral construction of frame and body. Apart from this novel feature, the Lambda also pioneered the use of independent front-wheel suspension in large-scale manufacture, the design in question being along lines first developed by Morgan, at Malvern, before the First World War.*

the next decade that designers began to take serious notice of the Lanchester findings. They were compelled to do so because the systems of independent front-wheel suspension under development and the kind of flexible frame tolerated until then were quite incompatible.

This problem had been observed at least ten years earlier by Lancia, an Italian manufacturer at all times open-minded in his attitudes towards design. His new Lambda of 1922 [116] had a rigid chassis of welded sheet-steel that formed the main structure of the body as well; coincidentally it had independent front-wheel suspension (i.f.s.). It set standards of good road behaviour for touring cars perhaps unequalled, until the introduction of the front-wheel-drive, integral-construction Citroën in 1934.

In 1913 Lagonda had introduced an 1100 cc car with combined body and frame, the panels of the former being secured to the side-members of the latter in a manner dictated by the economics of production and not by any niceties of design, cheapness being of overriding importance. In its way it was an elementary form of integral construction that was revived in the post-war 11.9 hp Lagonda model. The example was not copied; instead the industry chose to stick to traditional methods that allowed great flexibility of choice of bodywork to the all-important buyer.

The universal trend towards simpler body shape that was becoming evident by 1912 had continued to develop, and if it went a shade too far during the first year or two of peace a satisfactory balance was achieved by the mid-'twenties [XIII]. It had been speeded to some extent by intensive wartime development of the aeroplane, the forms of which had been greatly simplified and improved in a never-ending search for greater aerodynamic efficiency. This was reflected, a year or two after the Armistice, in the

117 *The chassis of 22-YG, a long wheelbase 40/50 Rolls-Royce, cost £2,000 in 1923. Its limousine body, of uninteresting shape, was built by Windovers Ltd, of Long Acre: it was finished in dark green and upholstered in light cloth, and it cost £733/12/–d. Extras, including a cobra horn (£6/6/6d), a Rolls-Royce mascot (£4), Lucas lamps (£27) and an AT speedometer and AT clock (£18/17/–d) amounted to £86/9/3d. Delivery was immediate in January 1923.*

'teardrop' shapes of such advanced cars of their time as the Rumpler from Germany, and the English North-Lucas, which was influenced by it. Car bodyframe manufacturing techniques also improved as a result of wartime developments in the use of jigs for high-volume production of aircraft fuselage and wing component parts in wood; standards of fit and finish were raised, along with rates of manufacture, and this state of affairs was reached at the best possible time, when demand was on the increase after the war.

For a time certain aeroplane features were adopted by car bodybuilders (the use of bracing wires inside doors, for example, to maintain rigidity, or the fitting of 'aero-screens', to cut down wind resistance), but marine influences on automobile appearance were possibly stronger; in any case they reached farther back, to the end of last century, when the recent introduction of the steam turbine had forced the evolution of new and more efficient hull shapes to take best advantage of its possibilities for speed. Some motor manufacturers had taken note, and action. The reader is referred back to the Leon-Bollée cars shown in plates 14 and 15, for examples. At the 1913 Paris Salon certain leading French coachbuilders showed bodywork unashamedly based on marine shapes (the Labourdette skiff was one such), and making use of marine constructional techniques as well. Even earlier there had been bodies like that fitted to the large Daimler shown in plate 66, and until the end of the vintage period, miniature ships' ventilators featured on many English sports-cars, as may be seen in plate 130, for example.

However, many early post-war cars had a mean look: their proportions were pinched, as it were; their curves were somewhat indeterminate [117], neither subtle nor voluptuously full-blown, and when they were deliberately featured, as

118 *The Grosvenor Carriage Co made a handsome job in this case. The car is a 1924 model 14/40 Vauxhall and its body is a folding-head coupé, with 'dickey' seat. Apart from its pleasing shape it is very well-finished, the fit of the hood, for example, being first class.*

119 *A justly famous American coachbuilder was Brewster & Co, an old-established concern, acquired by Rolls-Royce of America Inc. in 1926, which both before and after that event made superb bodies. The car illustrated is a 1925 40/50 hp American-made Rolls-Royce, with straight-lined but extremely handsome Brewster coachwork. On external evidence, at least, it compares with the best European work.*

occasionally happened, with closed bodywork, the result was seldom right. It was a bad period in appearance—because it was a time of transition, even if not all bodies looked so ill [118 and 119]. A similar state of affairs was to recur in the second half of the 'thirties.

In the early 'twenties the professional designer was poised to enter the American automobile industry. His concern was the outward show, the dressing of the product in as attractive a manner as possible, the stimulation of demand in response to a new policy of annual model change, initiated by General Motors soon after its direction passed into the hands of such able men as Pierre du Pont and Alfred Sloan. The latter summed it up succinctly: 'Fashion, not function, must be the determining factor.' Events quickly proved the commercial correctness of this dictum. Those car manufacturers who chose to stand still, to resist engineering change and to ignore the ever-increasing importance of appearance as, perhaps, the most essential selling feature, soon discovered that their sales graphs stopped rising. The classic case was that of the Ford 'T'—a long time a-dying because its practical value was so manifest.

120 *Walter Chrysler was another car manufacturer with an eye for good appearance; in his way, an American equivalent of Cecil Kimber or Sir William Lyons, for his automobiles combined above-average appearance with above-average performance. This 1926 Chrysler is typical of the medium-price touring car of its time, with side-screens and hood, a tool box on the wide running board, wood artillery wheels (with detachable rims), the kind of drum-shaped headlamps popular in North America and large bumpers.*

Nevertheless, the prolongation of its production life enabled its closest rival, Chevrolet, which had been committed to the Sloan doctrine almost from the start, to poach more and more of Ford's former customers. Certainly the introduction of the 6-cylinder engine was a factor in Chevrolet's ultimate success; essentially, however, the battle was won on style. This was a matter that Walter Chrysler always kept firmly in mind [120]. So, also, did William Brewster, at the top of the American custom coachbuilders [119].

The point was not missed in Europe. When William Morris, in association with the Edward G. Budd Manufacturing Company and another American concern, the banking house of J. Henry Schroeder & Co, established the Pressed Steel Company of Great Britain Ltd, in 1926, it was not only transatlantic ideas on body construction that came across the ocean. Early products of the new Cowley plant had a decidedly American look, externally, at any rate, and although there was always to be a difference in size between British and other European cars and their contemporaries from Detroit, styling similarities persisted until the late 'fifties, when Italian influences began to predominate. At last the traffic in design ideas reversed direction.

Thirty years or so earlier national characteristics were much more readily recognisable [XIV]. The individuality of appearance of a Rolls-Royce or Daimler, on the one hand, and of a 'Bullnose' Morris, AC [121], or tiny Austin Seven, on the other, was essentially English. The look of a 5CV Citroën, a sporting Delage or a Renault of any size or type [122] with 'coal-scuttle' bonnet, could only have orginated in France. A Mercedes of the early 'twenties, with its proud, pointed radiator and high-sided body, was unmistakably German, owing something to its forebears, the handsome, highly efficient racing machines that won the 1914 French Grand Prix; and Italian cars, almost regardless of make, had an honesty of design, a subtlety of line and proportion that set them apart [123 and 124]. The highest standards were reached between, say, 1928 and 1934, by Alfa-Romeo, the small-capacity sports-cars made by that firm (the 1½-litre models in particular) ranking with the best from any other country and, like them, richly deserving consideration as examples of high industrial design. Certain Bugattis had the same excellence of overall appearance, the type 35 in racing trim, or the type 51, for example, have the timeless look characteristic of man-made products of the highest quality. Rolls-Royce and Bentley, of English makers, provided splendid-looking chassis throughout the decade, fine foundations upon which the leading

121 *Both AC models of the early vintage years—from 1920 to 1927, say—were cars of strong individuality and good appearance—especially the 6-cylinder seen here at the John O'Groats start of the 1924 Monte Carlo rally. It had plain disc-wheels that suited the car (below).*

122 *The combination of Renault 'coal-scuttle' bonnet and radiator behind the engine, and bodywork as imaginatively and cheerfully decorated as this is distinctive and memorable. It dates from 1922.*

123 *Many Italian cars were bodied in England and some in Scotland. The Fiat illustrated, with Hooper landaulette coachwork of smooth shape and fine finish, is a 1923 model.*

124 *Two or three years later Hooper built this barrel-sided tourer on an Isotta-Fraschini chassis. The line curving down the side of the scuttle is not an aesthetic success and the shape of the rear is somewhat abrupt; nevertheless it is a strikingly handsome large car.*

121

GRANDS MAGASINS
AUX
GALERIES LAFAYETTE
BOUL^D HAUSSMANN-RUE LAFAYETTE

122

HOOPER & CO

123

HOOPER & CO

124

125 *An international mixture—German chassis, English body, Indian customer. Barker & Co built this aggressive looking coupé on a short-chassis 38/250 Mercedes-Benz for His Highness The Nawab of Bahawalpur.*

126 *A very formal town car—in fact a landaulette—built by Hooper on an 8-cylinder Packard chassis. It is a body abounding in subtleties of line and shape.*

coachbuilders of the time could practice their craft to the utmost. The great London houses, Barker, Hooper and H. J. Mulliner chief among them, with Gurney Nutting, Park Ward, Thrupp and Maberley and James Young not far behind, were busily occupied throughout this period, and not only with bodies for those chassis; other home manufacturers, as well as many from abroad, made use of their special skills [125 and 126]. By then their reputations were worldwide. They maintained the tradition of a centuries-old craft so far as attention to appearance, workmanship and finish were concerned, but the majority clung to traditionally conservative attitudes towards constructional methods and this was also true of the rest of the British coachbuilding industry. From the point of view of the progressive chassis manufacturer this was a disadvantage, for the average bodybuilder still equated strength with a lavish use of materials, and trying to persuade him that it might be gained more efficiently, more effectively and, perhaps, more economically by new approaches to the technical problems involved, was no easy matter.

It was the willingness of the London firm of Park Ward, itself new to the industry (but directed by men who combined great practical experience with unusual breadth of outlook) to tackle new constructional methods, that took them quickly into a close and enduring association with Rolls-Royce. Their first important collaboration was a scheme, initiated by H. I. F. Evernden, for the production of a range of standardised body types for the Twenty as early as 1922. Light-weight bodywork was an important design feature.

Bentley was another newcomer, his first car, the 3-litre, not appearing in production form until 1921. As a sporting machine it was best suited to open coachwork [127]; early examples, on large-diameter, narrow-section tyres, had a lean and workmanlike look but by mid-decade this model, with standard Vanden Plas four-seat touring body, was already a classic—that is to say 'of allowed excellence'. It was one of the best-looking English sports-cars of its day; well-proportioned, with a high and impressive radiator and bonnet, low of waistline, its narrow fabric- or metal-covered body placed well within the rear wheels and offering little air resistance. It was a style of body derived from much larger examples first seen on London/Edinburgh Rolls-Royce chassis between 1912 and 1914 and on the special chassis prepared for the 1913 and 1914 Alpine Trials; long, low and narrow; enormously elegant and impressive. The 1921 Austro-Daimler [xv] is of more ferocious aspect.

127 *Not all 3-litre Bentleys received bodywork of sporting aspect. Park Ward were responsible for this wide-winged tourer, which runs on narrow-section, beaded-edge tyres.*

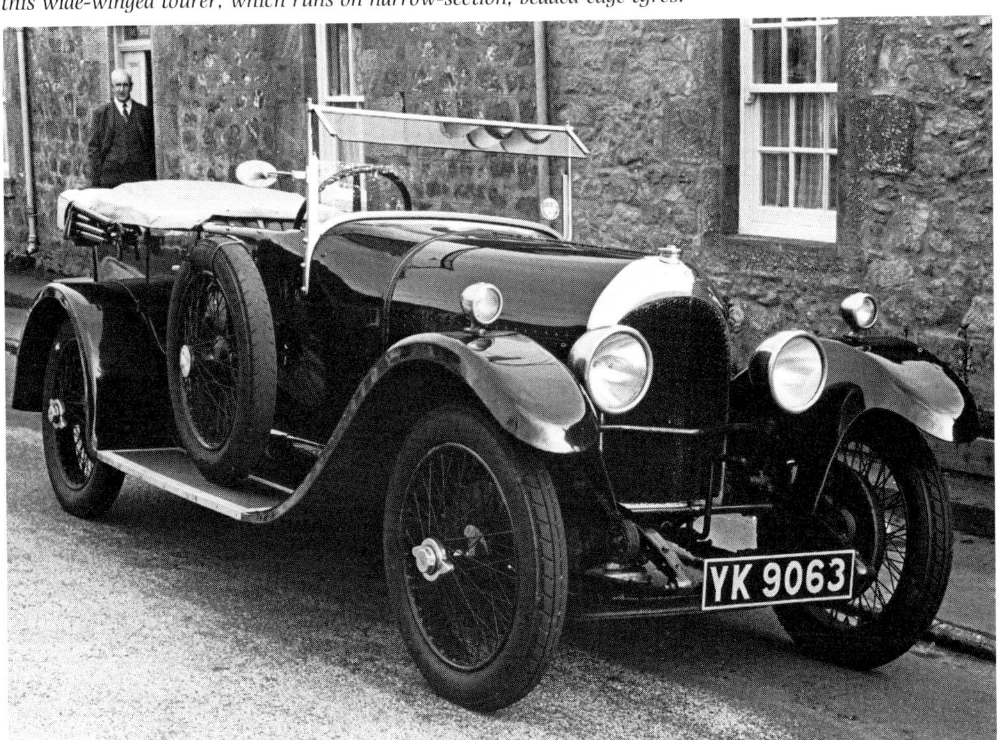

128 *A 1926 3-litre Sunbeam with its maker's standard open body. Its cycle-type wings are secured to each wheel individually and move with them. The cutaway in the body side allows the driver more elbow room.*

Although some may disagree, the writer feels that there are close affinities between those high performance luxury cars, sports-cars of the 'twenties such as the 3-litre Bentley and the original 1951 H. J. Mulliner Bentley Continental. These were four-seaters as a rule and luggage capacity was limited; they were not overbodied and not overweight; they, therefore had ample power while their engines were never stressed. Above all they had a magnificence of appearance that has no present-day equivalent.

Another good-looking fast car of the time was the 3-litre Sunbeam [128]. Its makers had taken an active part in racing before the war and resumed this interest after it was over, with unsuccessful appearances in the 1921 and 1922 French Grand Prix, a win in the Isle of Man Tourist Trophy Race in 1922 and, in the 1923 French Grand Prix, first, second and fourth places, no less. In 1924 Segrave, in a Sunbeam, was first in the San Sebastian Grand Prix and that was the end, more or less, of active Company participation in racing. In 1925 a 3-litre road car was second in the twenty-four hour race at Le Mans, and at later dates other 3-litres had some success at Brooklands.

This Sunbeam, like the 40/50 Napier, had a particularly fine engine. It was partly based on that of the successful 1923 Grand Prix-winning unit and had six cylinders, two overhead camshafts, two carburetters and an eight-bearing crankshaft. In spite of its high power output and performance, it ran smoothly, quietly and economically, a fuel consumption of 17 mpg being possible. It was a handsome, well-finished engine of the kind that is is a pleasure to polish. Unfortunately it was set into a chassis not quite worthy of it; cantilever rear springs might be satisfactory for a weighty luxury car but they were not really suitable for a 90 mph sporting machine. One hesitates to advance the opinion that their use was a cost-cutting measure, for the chassis price alone was £950, a vast deal of money by any standards then. The 3-litre Sunbeam was slim, neat and unobtrusive, seen from the front, but its cycle-type wings, the strong horizontal emphasis of its long and rather shallow bonnet and body in profile and the matching angles of its steering wheel and rear-mounted spares suggested something out of the ordinary. The restrained appearance was

129 *For the customer who wanted something special, Sunbeam were able to oblige. Unusually interesting features of this 3-litre are its polished (possibly clear-varnished) aluminium body, its long, pointed tail, with wood decking, its remarkable lighting arrangements and its combined running boards and tool boxes shaped like seaplane floats.*

characteristic of this make; Sunbeam's bodybuilding department was one of the best in the British motor industry [129] and nothing could be more English than a Sunbeam of the vintage period yet its designer, Louis Coatalen, was of French origin and was known to have sought inspiration from France and Italy.

The other significant sporting-car of the larger kind was the 30/98 Vauxhall, which was utterly English. The first post-war model, the side-valve 'E' type, was a handsome car [130]: fine-lined, subtle of shape, proud and distinctive at rest, audible and exciting in motion, to onlookers and occupants alike. In the manner of other cars of the period, it was high-set, standing on tall wire wheels and narrow-section

130 *The 'E' type 30/98 Vauxhall is a motoring classic; one of the greatest sports-cars of all time; tractable, comfortable and outstandingly handsome. This is a 1921 model, complete with torpedo-shaped sidelamps and ship-pattern scuttle ventilators.*

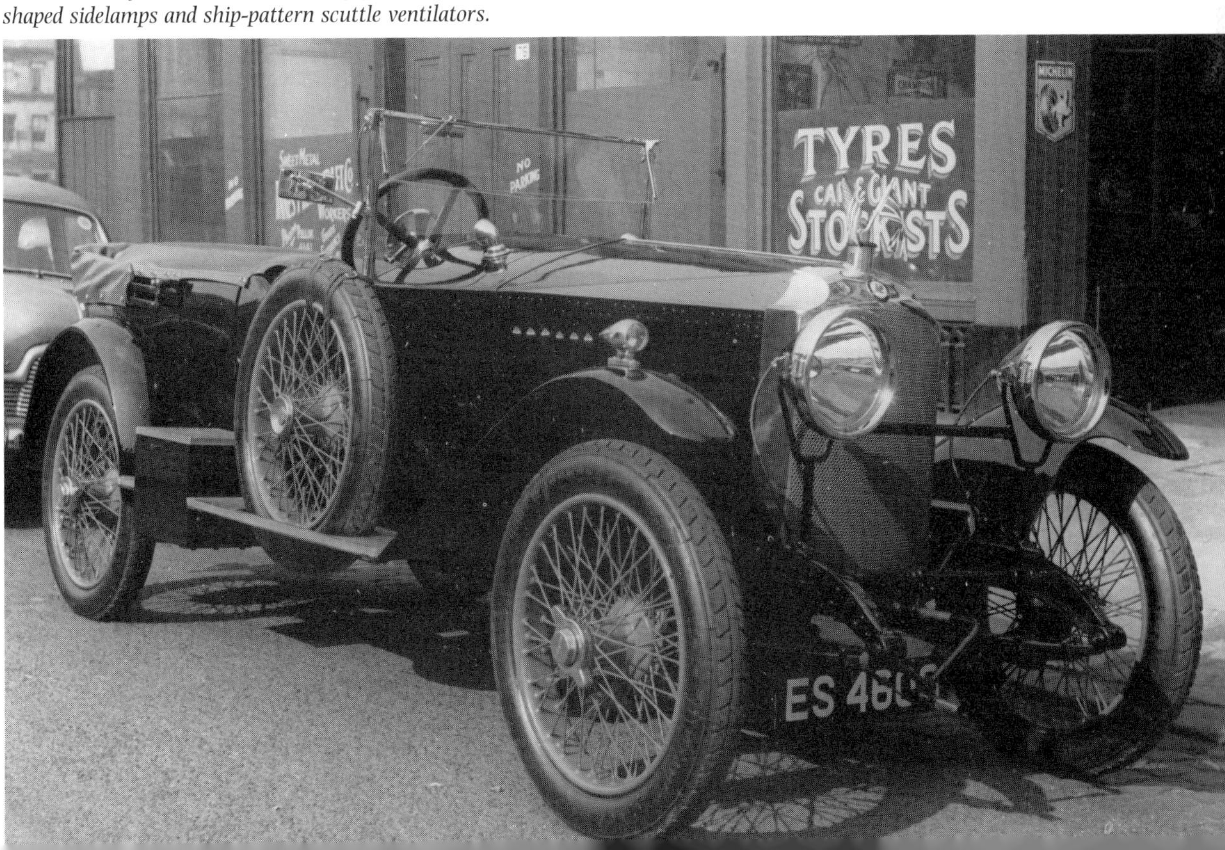

beaded-edge tyres, well clear of the road. Because of this and because its big, slow-turning engine had much torque it was a capable cross-country performer. Not that many owners chose it in the first place on that account; the 30/98 was sold as a fast touring car and that it most certainly was, 80–85 mph being within its capabilities. But rear-wheel braking made use of its full range of performance a matter of discretion, and even when four-wheel braking was introduced on the 'OE' model in 1923 the system chosen was not as effective as it might have been.

A skilled and intelligent driver has always been able to deal with problems of such a kind, however, and on the relatively empty roads of the early 'twenties he could savour the delights and excitements of travel to the utmost in one of these splendid cars. The writer has been fortunate enough to drive one 'E' type and two of the later overhead-valve 'OE' cars and can think of few more enjoyable experiences in a now-lengthy motoring life. Although these machines are large they are not particularly heavy and this is reflected in the unexpected lightness of their controls. They are fast but they are not temperamental like some speedy cars of their time, and flexibility is a noticeable feature; top gear running for most of the time is the order of the day, yet changing speed is a quick, positive and easy matter. So, too, is changing direction and, because of its beaded-edge tyres and lighter weight, the earlier model shows to advantage, there. In each, however, one is enormously impressed by the lack of effort, by the very long stride of these big cars, and by their comfort. By later standards the overall noise level might be thought a trifle high; but other factors more than compensate and it is only after a drive of 200 miles and more that it becomes intrusive.

131 *Another classic sports-car of the 'twenties was the $4\frac{1}{2}$-litre Bentley, seen here with a special short Vanden Plas sports touring body that is metal-panelled.*

The same can be said of most fast cars of that time but there was one that combined high speed with mechanical calm, and that was the $6\frac{1}{2}$-litre Bentley. It had a smaller engine than the 40/50 Phantom I Rolls-Royce with which it was in direct competition, it was as fast, or faster, it had a more usable gearbox and superior acceleration, and it was almost as quiet. On merit it ought to have sold very well indeed; it did not, relatively speaking, perhaps because the Bentley name was so closely associated in the minds of many with speed, competition and noise. The 3-litre was not inaudible, nor was the later $4\frac{1}{2}$-litre car [131]. But it must not be thought that only top grade cars offered driving satisfaction. At every motoring level there were models of especial interest or quality, and individuality [132], was much more marked at a time when design was in the hands of one person or of a very small team. The differences between cars of comparable specification, size and price could be very wide indeed, which is scarcely the case nowadays.

For obvious reasons, simplicity was a prime feature in the cheapest cars [133]. Materials were sparingly used wherever possible; there was no 'fat', no excess of wood, metal, glass or trim, and the latter was seldom leather; substitutes had been

132 *The Jowett was a simple, sturdy and highly-individual little car, built in Yorkshire, in hilly country, and it could climb almost anything. Its twin-cylinder engine had a characteristic thump and, like the contemporary Trojan, the Jowett appealed most of all to those of nonconformist taste (below, left).*

133 *The 10 hp Swift was a light-car of no great technical merit but it was simple and strong, it was relatively cheap and it was reliable. The rear view of a 1927 two-seater reveals the shape of its little body, which has much subtlety (below, right).*

134 *The Austin Seven was neat, relatively nimble and very well-made, despite its low first cost. One method of reducing its cost was to combine head and sidelights. The ingenious arrangement of its side-curtains provided a good shelter from cool winds while allowing access to proper open-air motoring.*

available for some time, the proprietary name, 'Rexine', being already well-known in the trade, for example; to describe a material similar to American cloth. The Austin Seven [134] was a good example of the ingenious ways in which the vintage-period designer and manufacturer made economies. It was a very small vehicle but its passenger and driver space formed the greater part. In general shape its little body was reminiscent of the domestic bath (but rather wider) and it, too, had an easily cleaned, stove-enamelled finish. Its interior furnishings consisted of two necessarily small, adjustable and removable bucket-seats and a shallow rear seat with a cushion wide enough to accommodate a number of infant posteriors.

This was functional design of a high standard. There was no attempt to disguise the fact that the Seven was a self-propelled vehicle; parts of its mechanism were clearly visible in the driving compartment, and, in any case, those out of sight were certainly not out of earshot; tiny the engine of the Seven might be but its growling and occasional tinkling accompaniments were an ever-present reminder of power production up-front. Although the model has always been called the first real miniature motor-car one is inclined to question this; surely such charming and practical little vehicles as the pre-war Calcott or 9.5 hp Standard, for example, have prior claims. But they were, however, scaled-down versions of larger types whereas the little Austin was an original, with no larger equivalents. Passing mention must also be made of the Bugatti-designed Peugeot Bébé which first appeared in 1913 and was certainly known to Austin.

The Austin was the right car, at the right time, however. In 1923, the first year of production, it cost £165 in standard touring form. The Ford 'T' tourer then cost £152 and the cheapest Morris, the Cowley two-seater, £299 5s. These larger cars cost more to tax and to run, but on the other hand they were roomier, faster and more comfortable. They were for the established family man and, along with contemporaries of similar type, served him well. The Austin was a kind of motoring *hors-d'oeuvre*; ideal for the man making a beginning with motoring and matrimony. It was not so satisfactory from the motoring connoisseur's point of view; driving a Seven, even a quite new one, involved too many variable factors for total peace of mind. The layout of the suspension system introduced a random element into steering that made preservation of a straight course on rough roads a matter of chance. Braking, too, was uncertain at best, the presence of a brake-drum on each wheel being more a matter of show than anything else.

Not so with the contemporary Morris cars [135]. Their roadworthiness had been proved over the years; they went where they were pointed and they stopped when directed to do so. They, too, were very well made and none would match the little Austin in terms of reliability. Again, simplicity of mechanism and bodywork were outstanding features. In one way there was nothing out of the ordinary in the general design of these cars, nothing that could be called unorthodox. However, the ordinary was transformed into the extraordinary by a philosophy that considered honesty of design, fitness for purpose, the use of good materials, high standards of assembly and an after-sales service and repair system on a country-wide scale essential elements. Needless to say other manufacturers followed suit. Reference has been made to the quietness of the Bullnose engine, the design of which was based on that of the American Continental units used in the original Cowleys and shipped over the Atlantic in wartime in considerable numbers. Its refinement of operation was matched by the sweetness of the clutch, which ran in oil and could last for up to 50,000 miles, if sympathetically treated. It was an important feature of these cars for it added much to the ease of driving, making changes of gear or take-off far less difficult and frightening than they could be on some other cars. In contrast, clutch operation on the Austin Seven, for example, required almost divine intervention.

135 *The 'Bullnose' Morris had a friendly appearance; it was workmanlike, with no pretensions to elegance, but it had strong visual appeal. A Cowley 'Chummy' two-seater is seen with another best-seller of more recent times, a Volkswagen 'Beetle'.*

136 *The instrument panel of a 'Bullnose' Morris Oxford is severely functional as long as it is not highly polished; then it serves as a reflector, or mirror (above, left).*

137 *The driver of the large 38/250 Mercedes-Benz is generously provided with instruments, some of which seem badly placed. All else is dwarfed by the huge revolution counter in the middle of a crowded fascia. The white steering wheel adds distinction (above, right).*

138 *An interior as lavish as this is a matter of taste and money. It is possible that modifications have been carried out since 1924, when the car was built. Perhaps not, however; American ideas on instrument panel design have always differed from ours. The car is a Springfield Rolls-Royce 40/50, Brewster-bodied roadster.*

The Morris owner was well provided with information about the running of his car. Facing him, on a vertical panel that was either of wood (Cowley) or polished aluminium (Oxford) [136], were four instruments, all white-dialled, handsome and legible. A Smiths clock to the driver's right was hinged on one side so that it could be swung open for access to its winding- and hand-changing keys; many have survived to serve as accurate and attractive domestic timepieces. A Smiths speedometer occupied the corresponding place on the near-side where the spectacle of its slender hand quivering slowly past the 50 mph mark was a major excitement for motor-mad boys of the present writer's generation. An oil-gauge and an ammeter—the latter set into a neat Lucas electrical switch panel—completed the tally of dials. More expensive cars [137 and 138] had more impressive displays of information. The flat-radiator Morris cars that replaced the Bullnose range in 1926 were sisters under their skin, mechanically speaking; the changes made were to coachwork, in the main, and were absolutely essential from a marketing point of view, since the appearance of vehicles of this make had remained almost unaltered since 1915. The look of the new models owed nothing to chance; it had been carefully studied, from front dumbirons to tail-light, from running boards to roof-top ventilator, in a manner increasingly common in the United States but hitherto quite unknown in Britain.

It was the first example in the United Kingdom of design for large-scale body production on a long-term basis, with provision made at the planning stage for annual changes of appearance to stimulate demand—the Sloan doctrine in essence. The hidden hand of the industrial designer was now revealed, yet he was still as anonymous as everyone else involved in the look of the finished product. The new and rather self-conscious steel-bodied Morris cars were available in closed and open form, as before, but the days of the two-seater, the two-adult, two-child 'Chummy' and the four/five-seat tourer were almost over, and the saloon or fixed-head coupé (the latter in penny numbers only) predominated. For fine weather motoring the open car [XVI] had a great many advantages and, over the years, had been developed to a high degree of excellence by motor manufacturers and specialist coachbuilders alike [139]. In a climate as unstable and unreliable as Britain's however, the

139 *One of the last genuine open tourers was made by the Humber company, which maintained high standards of design and construction almost to the end of the vintage period. The V-windscreen was characteristic and its shape was repeated in the instrument panel below.*

140 *As the primacy of the open car was lost and the closed body became increasingly popular, much attention was given to the design and production of opening roofs. The type illustrated, on a Renault, is still favoured by some manufacturers.*

141 *Three forms of transport photographed at South Queensferry in 1929; the great Forth Railway Bridge is a background for the steam driven ferry and the then-new Buick saloon. Its pressed-steel body is a great advance on earlier, starker designs.*

practical advantages of the closed body seemed more and more irresistible with the approach of the 'thirties, the more so when safety-glass became generally available and the opening, or 'sunshine' roof gained popularity [140].

William Morris was a practical man and a realist. He was quick to realise the potential of the assembled car and quick to exploit it commercially. Like any sensible European manufacturer he kept a close watch on what the American automobile industry was doing, and was also quick to realise the inevitable predominance of the all-steel closed body. In Europe his action in initiating the establishment of a pressed-steel body-plant was only anticipated by another follower and admirer of American methods, André Citroën, who re-entered the French motor industry in 1919 with a most carefully planned, volume-production operation that was almost instantaneously successful. Soon Renault had to follow suit and cars of that long-established make, with up-to-date all-steel bodies but traditional coal-scuttle bonnets, with a radiator-behind-engine layout, also Edwardian in origin, struck a strangely archaic note. Gradually, throughout the car-producing countries of Europe, the steel saloon took over [141 and XVII]. In a little more than a decade a very great deal had been learned about the properties of the material itself and about the techniques of its manipulation in the great power-presses of an expanding industry. Improvements in the steels used and improved methods of die-making, for example, enabled designers to evolve shapes of greater complexity, and between 1928, say, and 1934, much pressed-steel bodywork was notably well-proportioned, well-curved and handsome. Because of its transatlantic origins it is not surprising that it reflected American fashions of that period.

There were great improvements in external finish as well. For all but the most expensive kinds of work brush-painting had been superseded before the war, when the sudden shortage of skilled operatives in America caused a production crisis in

xv *The artist responsible has effectively caught the look of a large, fast, sporting-car, partly by exaggeration and partly by great skill. This is a 1921 Austro-Daimler.*

XVI *A 'twenties idyll—Simson-Supra sports-car, two pretty 'flappers' of sound social background and a fine day for open-air motoring. What a splendid evocation of the vintage years.*

XVII *By 1925 the future of the closed car was certain; by 1930 its supremacy over the tourer was confirmed. Here, in the earlier year, the artist has shown the two types side-by-side, each Buick characteristically high-wheeled and impressive.*

XVIII *The Leyland Lion was a widely-used single-deck bus in the late 1920s. Its design was a pointer to the way the bus was going—a lower, quieter, more comfortable and better-looking vehicle. This is a cast-iron model, by Wallwork, of Manchester, supplied to operators as a bus-office window attraction.*

1909. It had also speeded up the introduction of the practical spray-gun. In 1925 there was further progress as far as the finishing of volume-production bodies was concerned when cellulose was introduced, and so good was this new, easily handled and easily maintained material that within a year it was taking the place of traditional paints and varnishes. A little later, around 1928, chromium-plating appeared and another motorist's chore was reduced from hours of arduous toil (as in the case of polishing the nickel-plated parts of a motor-car) to a minute or two of easy work with a damp cloth. Later on, stainless steel was to provide an even more attractive alternative to the admittedly lovely colour and fine finish of good nickel-plating. The latter was splendid for those with the time and energy to maintain it in top condition. For the majority, however, the colder, harder look of chromium was of little consequence. It was practical, and that was what mattered to an ever-increasing number of motorists for whom the automobile was no longer a novelty, a toy to be fussed over or tinkered with at weekends, but a utility [142], an aid in business and work or play of every conceivable kind. As far as Europe and much of the rest of the world was concerned, there was an increasing acceptance of the motor vehicle as a necessity of life and not a luxury: even so it was to be a long, long time before it was taken for granted in the North American way.

Between 1919 and 1929 the motoring scene in Great Britain changed enormously. The number of vehicles in use rose from a little over 200,000 to close on 1,500,000 in ten years. In 1919 there were 62,000 goods- and 44,000 passenger-carrying vehicles in use; for 1929 the corresponding numbers were 336,000 and

142 *A typical English road scene in summertime in the late 'twenties. The big sleeve-valve Daimlers and the Rolls-Royce are hire cars; in the distance can be discerned a Lancia Lambda saloon.*

99,000. There were 110,000 cars on the road in 1919; by 1929 that total had soared to 998,000, a ninefold increase. The whole life of the country was transformed.

For those who could afford it the motor-car provided almost unlimited mobility. For those who could not, the motor-bus was a very good alternative: it was cheap, reliable, frequent, and although comforts were minimal in the earliest post-war models, with their solid tyres and wooden-slatted seating, there was a steady improvement through the decade. Long before 1929 in fact, standards of comfort on public transport approached, and sometimes equalled, those enjoyed by car owners. Apart from local service operations, long-distance journeys were undertaken as a matter of course, much to the annoyance of the railway companies, which were also affected on an increasing scale by the growth of goods transport by road. Where the train could carry goods from door to door, so to speak, it still held all the advantages, but that state of affairs seldom obtained; not every pit-head, after all, had a direct rail-link to a consumer. Although the railways retaliated by operating their own fleets of 'feeder' vehicles to transfer goods from train to ultimate destination, provided road-rail vans particularly suited to household-goods transport and maintained useful local parcels services, the general trend was towards greater use of road transport. The industry also had a powerful lobby in parliament. Its vehicles were ideally suited to the carriage of small loads but, as mechanical efficiency increased, and as better braking and the pneumatic tyre made the carrying of heavier loads a safer, quieter matter, the scope of the 'commercial' gradually broadened. One way of dealing with an exceptionally large load had been to divide it between lorry and trailer, and this was a common practice. However, it was wasteful of road space, for one thing; for another it provided drivers with all manner of additional worries and practical dangers.

The problem was solved, eventually, by the introduction of the articulated vehicle, Scammell being a pioneer make in this respect, concentrating on high-capacity

143 *With a hood of such a size, numerous side-curtains and five doors on each side to deal with, the builders of this Renault coach body had many problems to solve. It cost £875, had a 17.9 hp engine and seated twenty, side by side, on five rows of seats.*

units, its ultimate triumph being the design and manufacture of two 100-ton low-loaders of exceptional technical interest. For a long time, however, the more usual way of achieving extra load-carrying capacity was to add an axle—a practice also adopted for vehicles intended for off-road, cross-country use. In the latter case, both rear-axles were driven, for extra traction; for exclusive road use, however, the third axle was sometimes of the 'dead' or trailing type.

The rigid six-wheel chassis was also used for omnibuses for a time, for single- as well as double-deckers, the extra frame length making it possible to increase the number of fare-paying passengers. The type failed to gain as much success as might have been expected—though it remained popular with trolley-bus manufacturers until the end—and was quite out of fashion by the early 'thirties. One suspects that excessive tyre wear on the rear wheels was one reason; axle trouble, probably, another.

The open motor coach or charabanc [143], in effect a greatly enlarged touring car, remained in general use throughout the decade; it was popular for outings and touring in spite of the effort involved in handling gigantic hoods in unfavourable weather. Its body, which featured full-width bench seating, as a rule, and required a door at either side of each row for proper access, was quite costly to build in consequence. The service bus, for everyday use, was closed, of course. At first it stood far above the road [XVIII] since adapted goods vehicle chassis were used, with high-set frames: but as early as 1923 the London General introduced a lower chassis for its NS type, and towards the end of the decade there was a general lowering of chassis and

144 *Sir Dennistoun Burney's rear-engined streamlined saloon went into small-scale production in 1930, and although the Prince of Wales was a well-publicised purchaser, the unconventional design of the car, its high price and the growing slump made its commercial failure a certainty. The low-set 1928 prototype was a good deal uglier than the production model. It was based on a front-wheel drive Alvis chassis turned back-to-front.*

bodies improving appearance and making access easier. At the same time, the motor-car was getting closer to the ground [144]; the days of the walk-in rear compartment, as seen at its most extreme in the giant Royal Daimler State cars, were almost over. The closed motor coach was also much liked for excursions and for long-distance work, where its high standards of amenity were greatly appreciated. Saloon-car comfort and

145 *A GMC truck chassis which was also sold for bus work; a representative American product of the late 'twenties with fuel tank below the driver's seat and instruments that were small and difficult to read.*

quietness were matched, without doubt, by these fine petrol-engined buses. The coming of the oil engine cut operating costs dramatically but raised overall noise levels to an almost unacceptable degree.

By the beginning of the 'thirties, goods vehicles were a good deal easier to drive than they had been ten years before. The cast, spoked wheel and the solid tyre had all but disappeared. There had been a steady rise in mechanical efficiency: engines were more powerful, more reliable on the whole, and longer lasting; transmission systems were better, though unchanged in essentials; braking was about to make a universal change-over from rear-wheels and transmission to all four wheels, with hydraulic operation already available and servo-assistance increasing in its applications. Bodywork was no longer stark and ugly; where appropriate it was well-rounded. Cab facilities were very much better—one small but important point was the fitting of a speedometer as standard equipment, and close attention was paid to methods of achieving weight reduction without loss of structural strength.

As far as the market for the larger kind of goods vehicle was concerned, British manufacturers were the principal suppliers. Ford and Morris held a major share of the market for vehicles up to one-ton capacity but American makes were popular: Chevrolet, Dodge and GMC [145], for example, and French and Italian models had their own following. A similar state of affairs prevailed with bus operators, who had a wide choice of vehicles of the smaller kind from the same sources, home manufacturers concentrating on full-size single- and double-deckers and taking surprisingly little note of a rising demand for twenty-seaters until a late stage of the period. In 1929 imports of completed cars, goods vehicles and buses, and of chassis, amounted to 37,784; exports to a neatly-rounded 42,000. The latter was a great improvement on the 1922 figure of 3,041 but it represented little more than one-fifth of actual production. Many years were to pass before that percentage showed improvement.

By and large the British car was best suited to home conditions. A taxation system that imposed a duty of £1 per horsepower (the latter calculated from the outdated RAC formula), discouraged the production of engines much above 13.9 hp, the resultant market gap being filled here, as well as overseas, with low-cost, high-power, go-anywhere American cars. It encouraged the perpetuation of the tax-dodging, long-stroke engine and the introduction of Nines and Tens to tempt those whose initial motoring experience might have been gained with the ubiquitous Austin Seven. By 1929 Humber, Riley, Rover, Standard, Singer [146] and Swift were prominent in that class. By far the best and most interesting was the Riley Nine [147], a model that had been in production since 1926. It had a high-efficiency but untemperamental engine with twin, high-set camshafts and great reserves for development; in other words its power output could be substantially increased with no loss of reliability. The chassis was low-set and so, too, was the body, which cradled its occupants between the axles and, in general layout and appearance, was to have a strong influence on other makes. The early close-coupled Monaco saloon was awkward-looking but improvements were so quickly put into effect that by 1930 it

146 *In a bodybuilding department as busy as that of the Singer company, (which was one of the largest English producers in the late 'twenties), much use was made of jigs and power tools to sustain a high level of output. Here the pre-machined members of a composite-construction saloon body are being drilled on a special fixture before assembly.*

147 *From the outside the Riley Nine Monaco is deceptive. The widening of its body in plan gives extra space for front-seat occupants, and at the rear there is ample height and leg room.*

was one of the most handsome and distinctive small cars on the road. The high quality of its appearance set it apart for the best part of a decade.

In what might loosely be called the middle-class of motor-car AC, Armstrong-Siddeley, Crossley, Daimler, Humber, Sunbeam [148], Talbot, Vauxhall and Wolseley were in competition. Technically speaking, the most original and interesting was the Talbot 14/45, designed by Georges Roesch, the Company's Swiss-born chief, and introduced in 1926. It had a very smooth 1,666 cc 6-cylinder engine of notably clean external appearance, excellent steering and most effective brakes, but little performance because it was underpowered. The 14/45 founded a line of historically important motor-cars of handsome appearance and great mechanical distinction.

It was in this class that easy gearchanging was first introduced to the British market. The Wilson-type pre-selector gearbox was adopted by Armstrong-Siddeley, then by Daimler, and undoubtedly eliminated what was still the most difficult part of driving for most people. In its Daimler application it was coupled to a fluid fly-wheel; in the hands of a really skilled driver this combination could scarcely be matched for smoothness of take-off and changes of speed. But another, simpler, less expensive aid to easy changing was on its way, from America.

Something has been said of the English sports-car already—but nothing of makes and models of such quality and appeal as, for example, the Alvis 12/50, the Aston-Martin $1\frac{1}{2}$-litre, the Frazer Nash, the $4\frac{1}{2}$-litre Invicta, the Lagonda 2-litre, the Lea-Francis 12/50, the 'M'-type MG (derived from the Morris Minor), the Riley Redwing and Brooklands. In performance and in handling they were vastly superior to the majority of contemporary touring cars, this being proved by the fact that so many have survived and are still in regular use.

The number of cars in the highest class was small by the late 'twenties. Bentley had established a great reputation, partly because of the quality of his cars and partly because of numerous successes in racing but, because of a chronic lack of capital, his firm was never able to take the share of this market that their excellence merited. Although its name was still highly respected, Daimler was finding it more and more difficult to maintain sales by the late 'twenties. Indeed, an appropriate slogan for that concern might have been 'Desperation, thy name is Daimler' for by 1927 twenty-three different models were catalogued, based on five distinctive chassis types ranging from the 6-cylinder 16/55, at £490, to the liner-like 50 hp V-12, at £1,950 in long wheelbase form. Bodywork was extra, of course; as made by Daimler it was of fine quality.

Lanchester still had the Twenty One and Forty models in production but the new Straight-Eight was gradually taking their place. Rolls-Royce outsold its rivals by a substantial margin and was able to maintain its leading place in spite of a recurrence of worldwide economic problems. Between 1919 and May 1925, when it was replaced by a new overhead-valve engined model, the Phantom I, about 3,000 40/50 chassis were built. Two thousand, two hundred and twelve Phantoms were made between 1925 and 1929 and 2,940 Twenties between 1922 and 1929. Thereafter there was a gradual drop in sales which reflected the difficulties faced by all makers of costly cars during the next decade.

148 *The front aspect of late vintage-period Sunbeams was most impressive. The amount of extra lighting equipment suggests that this car may have been a Monte Carlo rally entrant.*

149

150

151

5 *Motor-cars for all 1930 to 1940*

Everywhere sales of the more expensive kinds of car were hard-hit by the economic troubles that affected most countries after the Wall Street Stock Market Crash of 1929, but as far as Europe was concerned this happened gradually [149]. In the United States there was a rapid reduction in the number of those individual makes underway since the mid-'twenties, and many once-famous names quietly disappeared. In Britain, the slump did not take full effect until 1931, but even then, despite an appallingly high level of unemployment in certain areas, notably in the north, the situation was never quite so serious as it was across the Atlantic.

It would be incorrect to claim that the British motor industry remained unaffected, however. There were failures of individual companies; mergers, or take-overs, became necessary for the survival of others; some went into a slow, ultimately fatal, decline. Overall, though, the car, lorry and bus makers, the coachbuilders, the manufacturers of components and accessories and the suppliers of raw or finished materials, entered the 'thirties in slightly better order than might have been expected [150].

Because statistics can often give a more immediate picture of an actual state of affairs than words alone, no apology is offered for their renewed use. In 1923, for example, the British industry produced 71,396 private cars and 23,604 lorries and buses. In 1929 these totals had risen impressively, to 182,347 and 56,548, respectively. In 1930 cars dipped, to 169,669 [151], but lorries and buses were up

149 *Period scene at Ballachulish, in the Scottish Highlands, in 1930. An Austin 16 saloon, its roof-light open in the summer heat, sits on the little ferry, while hikers, flappers and locals gossip. The world slump seems remote.*

150 *Regardless of the economic state of the world, some drivers tried to go faster than others; some tried to drive farther than anyone else, and some were content to attempt the crossing of continents; the latter is happening here. A team of specially-designed, special-bodied Citroën-Kegresse half-track vehicles spent much of 1931 and 1932 traversing Asia, from Beyrouth to Peking, with crossings of the Himalayas and the Gobi Desert* en route.

151 *Although Vauxhall Motors were still small-scale car manufacturers in 1930, it was considered economical to operate a sizeable press-shop for the production of panels, mudguards and other body parts. By then, the various processes were largely mechanised but there was a fair amount of hand-finishing involved; a state of affairs maintained to the present time in volume production of bodywork.*

again, to 66,859. Cars fell to 158,997 in 1931; lorries and buses rose slightly, to 67,310. After that, apart from 1932, when the latter dropped to 61,475, all production totals rose steadily until 1937. In that record year 389,633 cars were built and 118,116 lorries and buses. In the following year, according to another statistic, there were 40,000 chauffeurs in the United Kingdom.

The motor vehicles of 1930 obviously differed greatly in appearance from those of 1919. Few still ran on solid tyres—the occasional Trojan among their number—and the skinny high-pressure tyre had been superseded by the low-pressure balloon during the second half of the previous decade. This change not only improved ride comfort: it gave cars so equipped a 'better-nourished' look. The fatter, heavier tyres brought difficulties with them, which principally affected faster cars. Wheel-tramp, or -wobble, was the worst, and it took some time to determine its causes and to provide a cure. A contributory factor was a certain loss of rigidity at the front end of the car, resulting from the introduction of flexible engine mountings.

Another problem was directional control but this, at least, was less potentially dangerous. The larger area of rubber that the balloon tyre offered to the road made steering heavier, and to counteract this, lower gearing was provided. In practice this meant that more movement of the steering wheel was required to move from lock to lock, a feature disliked by the keen and sensitive driver, who was prepared to tolerate some heaviness at lower speeds as long as there was no loss of quickness of response or precision. More and more, however, his was a minority voice. There was a universal move towards the production of vehicles that would be as easy as possible to manage, and lighter steering, was an important part of it, even if it did require more wheel turning.

To a considerable extent, the continuing rise in the number of women motorists put pressure on designers and manufacturers to make driving simple and physically undemanding, within the limits of available technical knowledge and resources. As might be expected, the owners of American cars were already in a privileged position; their large-engined automobiles ran in high gear most of the time anyway, and with only three speeds to deal with, as a rule, the amount of 'shifting' to be done was small. This also applied to the more powerful kinds of European car. As far as steering was concerned, again the transatlantic driver had been used to more wheel-twirling than was considered necessary or desirable over here.

For many motorists still the greatest problem was gearchanging. Although the whine or groan of gears in mesh was becoming less and less noticeable on the road (for which fact steady improvements in design, wheel-cutting and assembly were responsible), their clashing together was still frequently to be heard. To be fair, it must be said that some makers made it as easy as possible for the driver. The 'Bullnose' Morris change was a good one and quick to learn and, although it seemed very difficult at first, that of the Austin Twelve was most satisfying, once mastered. Even better to use were the Humber and Sunbeam gearboxes, which were especially light and precise, whilst surprisingly tolerant of bad timing. Until 1929 the 40/50 Rolls-Royces had a change that was far from easy for any stranger; before the gear-lever could even be moved it had to be cleared of its retaining-notch, and in a fluster of traffic it was not difficult to forget this vital fact. The smoothness and quietness of the engine, moreover, made it exceptionally difficult to determine its speed with a sufficient degree of accuracy for clean and silent shifts of speed. There were mysterious ways and means of changing gear on a Rolls-Royce, quietly and smoothly, as swiftly or as slowly as one wished, but these were only to be learned at its maker's unique Drivers' School in London.

Salvation was at hand, however. In the United States, in 1928, Cadillac had introduced synchromesh, a mechanical means of properly synchronising the speeds of meshing gearwheels during the change from one ratio to another. The system did not cross the Atlantic until 1931, when it was a feature of the smallest Vauxhall to appear since the early years of the century—the two-litre Cadet, which had been introduced in 1930. The much more distinguished 20/25 hp Rolls-Royce [156] was provided with synchromesh in 1932, on its third and top gears, and in the following year the gearbox of the lusty 40/50 hp Phantom II was similarly equipped [153].

By the mid-'thirties synchromesh was in general use, fitted most commonly to the two highest gears. There was a tendency thereafter to add it to second speed as well (in 1935 the Hillman Minx anticipated very much later practice by appearing with synchromesh on all its forward gears); of its usefulness there can be no doubt

152 *Good forward visibility is essential; it is a safety factor of the utmost importance. Well-placed and legible instruments also play a vital role in informing the driver of speed, state of engine temperature, oil pressure, fuel supply, rate of battery charge and so on. From the driver's seat in this Riley Nine all is easily seen. The shapely scuttle cowls serve as wind deflectors when the windscreen is folded flat. Note the elasticated, pleated map pockets on the door trims, the Jaeger speedometer, Smiths rev. counter and Lucas wiper-motor on the windscreen frame.*

153 *In spite of the length of its bonnet, forward visibility from this 1935 Rolls-Royce Phantom II is very good, the top of each front mudguard being well in view. Gear- and hand-brake-levers are tucked away on the driver's right. The extremely legible instruments are symmetrically arranged in the centre of the painted wood fascia of this particular car, which is an all-weather or convertible, with bodywork by Gurney Nutting.*

154 *While the high bonnet and tall radiator looked impressive enough from the outside they tended to cut down forward vision for those inside. Perhaps the owner of this 1930 Vauxhall 23/60 did not mind; he had, after all, an extremely smart two-seater, with 'dickey', made by the Grosvenor Carriage Co, a concern that handled this type of body with some style, as earlier examples prove. The paintwork is cellulose, the bright metal parts are chromium-plated, there is a locker just ahead of the rear mudguard shaped to house a golf bag and supports for another just in front of the 'dickey'.*

whatsoever, for it greatly reduced the difficulties of driving on the increasingly congested roads of the period. In this way, easier gearchanging added a certain element of safety to everyday driving.

There had been a 'Safety Stutz' in America in the late 'twenties, and the MG slogan for many years was 'Safety Fast'. 'Safety Glass' had been available since before the war, but it was not until 1927 that its use in cheaper cars became popular, and sometime later before it ceased to be an extra-cost item of equipment. Protective bumpers, or 'fenders' [XIX and XX], as they are known in the States, may not have been liked by Henry Royce but their utility, particularly in urban traffic, was undeniable.

Many mechanical features of the better sort of car, whether of the touring, sporting or luxury type, directly affected its safety in use. Good brakes, good steering and good roadholding were all positive factors in that respect. Good visibility for the driver was another and this feature, so widespread in the days of high seating and low bonnets, was to suffer during the 'thirties [152 and 153]. The tall radiator and high bonnet came into fashion [154] (often there was little enough beneath the latter), and the near-side wing—the one farthest away from the driver, that is—was invisible to him on more and more cars. The separately mounted side and headlamps were still standard on most motor-cars, and served as valuable markers, by day as well as by night. The fact that headlamps required a surprising amount of power to push them through the air was not general knowledge. Drivers who demanded the extra 5 mph of top speed knew what to do; either they removed the lamps altogether (which was not illegal) or turned them round on their mountings so that they presented their tapered ends to the front. Those large and beautiful lamps—in particular the Lucas P.100, the Marchal and the Zeiss—were most effective illuminators of a kilometer or two of road at night. The quality of materials and workmanship was unsurpassed. In 1928, a pair of Lucas P100 High-Power Mirror Projectors cost £22 with nickel-plated finish. With silver plating they cost £30. Their value, in visual terms, would be impossible to determine.

In the early 'thirties there were many handsome vehicles about [155, 156 and 157a]. The clumsy look of the cars of a decade earlier had gone, along with small-section tyre, the large-diameter steel or wood artillery-type wheel and the badly

155 *The appearance of the largest cars of the early 'thirties could be very striking, as in this example. The car is a Phantom II Rolls-Royce with what Barker & Co, its designers and builders called a saloon limousine body. It is, in fact, a sports-saloon of the close-coupled type developed for the 'Continental' 12 ft wheelbase Phantom II but with many distinctive features. The lavish display of louvres is a period conceit; those above the windows serve as anti-draught ventilators. The helmet-pattern wings, the waist moulding and the rain-gutters are finished in ivory, with lining of the same colour on the bonnet and body-side louvres. There is a Yale lock on each side of the bonnet and on all four doors. The steps below the latter have covers integral with the panelling, to keep them clean and dry in bad weather.*

156 *At this time, too, the small Rolls-Royce appeared with exceptionally handsome coachwork. This 1932 20/25 hp model has a superb four-light saloon body by H. J. Mulliner; a composite of aluminium panelling on a wood frame in traditional manner.*

157

157a

158

proportioned body. Formerly, far too many cars were badly dressed; their suits were badly cut, yet there was nothing wrong with the quality of their materials or their workmanship. There was a steady improvement from, say, 1924; one dare not be more precise because pace-setting examples from earlier years are always to be found. The austere look of some 'twenties cars could be compared, not unreasonably, with the grimmer aspects of the typical fictional spinster. For proof consider the angularities and minimal curves of the 1921 Wolseley tourer. By the late 'thirties one would have been making comparisons between Miss Mae West and, say, a Dodge sedan! A few years earlier, however, a proper balance had been struck and many cars of that time were splendid of their kind. For obvious reasons the smallest must be excluded; their wheels were too big for a satisfactory balance of proportions to be achieved, and body shapes were ungainly. Early Morris Minors were not at all good-looking [157]. The first MG Midget—basically a re-clothed Minor—was attractive, not handsome, and the Austin Seven saloons, in retrospect, were either comical or rather pinched and mean-looking.

Before long, however, Cecil Kimber had contrived to turn the smallest MG into a sports-car of distinctive style and appearance, the J2 and its successors superior in these respects to their contemporaries of similar size and type. Kimber was one of those rare men who made no fuss about his nose for aesthetic quality [158]; possibly he could not draw, in the conventional way, but he had a sense of style, the proof of which is to be seen in the look of the cars for which he was directly responsible over a period of fifteen years or so. They were individual in appearance, and original, too: Kimber's eye for the lean and elegant line had an Italian subtlety and inevitability, but was not derivative. He had a special ability to suggest rich forms without the use of ample curves. He was a sculptor, and a good one.

William Lyons was the other important English designer of the 'thirties (and for long afterwards). His well-rounded sidecar bodies and special coachwork for small, cheap cars began to attract favourable attention long before he moved from Blackpool to Coventry in 1928, to be closer to his customers and suppliers, in what was still the centre of the industry. His early work compelled attention. The Swallow-bodied Austin Seven saloon, for example, had individuality and a boldness of shape and colour that would not have been allowed at Longbridge, where visual reticence was the order of the day. The first cars for which Lyons was wholly responsible (as an

157 *The original Morris Minor was a badly proportioned vehicle, a fact that did not deter potential owners attracted by the maker's reputation and its known liveliness. The body was fabric-covered and very light; it was also easily and cheaply built. Bonnet and wings were steel and the windows slid open, as those of the Mini were to do thirty years later.*

157a *Despite certain visual affectations the overall shape of this 1930 American-built Rolls-Royce is positive and anticipates trends developed elsewhere at later times. It was a Brewster production (William Brewster named it Windblown); it was at the 1930 New York Auto Show; it was not a total practical success, and it still exists.*

158 *Under Cecil Kimber's direction, the shape of MG cars was always interesting; often extremely good. The basic layout of open two-seaters stemmed from the 1932 Midget, with its rather long bonnet, its cutaway doors and its tall, vertical petrol tank standing at the rear of the chassis, just ahead of the spare wheel.*

overall design concept, that is) were the SS1 and SS2–Standards beneath their rather self-conscious exteriors. For their price, their appearance verged on the sensational; the SS1, the larger model, seemed to be all bonnet and very little body, with everything so close to the ground, after the manner of the contemporary 4½-litre Invicta [159] or the special low-chassis Double-Six Daimler with which Laurence Pomeroy, Senior and Reid Railton had been involved a year or two earlier. If those first SS models did not establish the long, low bonnet fashion in the first place, they and their much better-looking successors certainly perpetuated it most successfully.

To launch a new make of car in 1931, when the economic slump was making things so difficult in so many other ways, took a great deal of courage. As the statistics have already shown, however, British car sales were only slightly affected by the crisis and William Lyons already knew that his particular ability to create exciting but acceptable bodywork was a marketable asset. By collaborating with Standard, an old-established concern noted for conservatism and for the reliability of its motor-cars, he was making sure of the mechanical side; as for the rest, his capabilities as a designer and builder of special bodywork were already widely respected. From the start it was clear that his recipe was a sound one: this particular dish was showy but it had a most distinctive flavour. The SS [160], in its early days, was not simply a Standard in fancy dress; it was a different kind of car altogether, with its own personality and its own attractions. Its performance, too, was of a higher order. In this respect it was much more honest than either the Austin Twelve Sports Tourer of a year or two later, or the rival Morris Ten-Six, which were little faster than their squarely sedate sisters.

Lyons eschewed the wire stoneguard on headlamp fronts and radiators, and the bonnet strap; he had greater visual resources at his command, and his second season's cars were proof of an ability to move forwards, always refining and improving, and not resting on any laurels. The mudguards of the first cars had been of the helmeted type then fashionable in some quarters. They looked something like the 'cycle' wings seen on the 1½-litre Aston-Martin [162a], but Lyons' mudguards did not move with the front wheels or rise and fall with the rear ones. For 1934 he introduced long front wings that flowed gracefully towards the rear; their line was firm and confident and emphasised the length of the car. His first open SS, the 90, was handsome and lively. For a year or so, in the middle of the decade, there was a fashion for the 'streamlined' saloon that had originated in France some years before. By the time the style reached Britain, however, most of the visual excitement had gone, and what was offered by Rover, Talbot and Triumph, for example, was neither aesthetically nor functionally very satisfactory. The SS Airline managed to look rather different from its rivals, with its long, sweeping wings and positive curves, and attracted much attention on the road. The interior, it is said, could become uncomfortably warm on occasion and rear-seat passengers almost had to be shoe-horned in. Relief was at hand, and one of the most important new models of 1935 was the SS Jaguar sports-saloon, available in 1½- and 2½-litre versions. Their bonnet lengths differed, the smaller model having a 4-cylinder engine, the larger a 6, but their bodies were identical, which was one of the reasons for their remarkably low prices. Although the new Jaguars were low in build there was plenty of headroom

159 *The $4\frac{1}{2}$-litre Invicta was certainly among the best-looking large English sports-cars of its day; long, low and truly handsome. This 1933 model is not at all flamboyant; its looks are functional and purposeful. Hood stowage is especially neat.*

160 *Within a year or two of entering the industry as a motor manufacturer, William (now Sir William) Lyons was building cars as distinctive in appearance as this SS2, the lower-powered model in his early range. It was a four-seater but those at the rear were somewhat cramped for leg room. It was a little ostentatious for some but a welcome relief from the generally sedate steel-bodied or coachbuilt cars in the under £300 class in 1933–34.*

161

162

162a

inside and plenty of width, and the overall appearance was first-class. Lyons handled curves with great confidence at all times, and, in the new cars he combined the old helmet shape with the flowing-profiled mudguard in masterly fashion. In side-elevation these cars gave the impression of pent-up energy and movement, and this was borne out by their performance, which was comparable with that of the Bentley.

Someday, perhaps, it will be possible to examine costings for those pre-war SS cars. In the meantime one has to wonder how it was possible to market a handsome, fully equipped four to five-seat saloon that could comfortably exceed 85 mph and in mechanical calm, moreover, for £385. And if that was impressive by 1935 standards, for a 2½-litre saloon of real distinction, it was as nothing in comparison with the kind of value for money offered by 1939, when the same model, refined and improved in a variety of ways and with an 'air-conditioning' system (the first fitted as standard to a British car), sold for £395. One size of saloon body continued to serve for the 1½-, 2½- and 3½-litre models, but by then a most attractive drophead coupé was also listed, for only £20 more, and 2½- [161] and 3½-litre versions of the 100, a two-seat sports-car of superb appearance. It was the best-looking vehicle of its type in 1939 and one of the very few classic designs to come out of a period more notable for poor standards of appearance design than for visual excellence.

Kimber's magic touch had not been adversely affected by the changes of policy and of direction in the Nuffield empire during the second half of the 'thirties, and the Midget in its last type, the 'T' series, was by far the best-looking of the small British sports-cars. The largest MG models then—the 2-litre [162], first seen in 1935, and the later 2.6—were something of a mechanical mixed-grill (like the Midget they used Wolseley engines, for example), but they were large and impressive machines, in direct competition with the Jaguar [161].

The other British make that had been much above average in the design and layout of its coachwork was the Riley; although the Nine was an awkward-looking car in its first year or two, there was a steady improvement, year by year. The standard Monaco sports-saloon had reached a most satisfactory level by 1931, after which it was only in detail that changes in equipment and gradual refinements of

161 *Like the Invicta and some other contemporaries, the SS100 had a long bonnet and a shorter space for driver and passenger, hood, fuel tank and spare wheel. It would be difficult to improve upon the way in which these parts were proportioned and combined by Lyons. It is said that he and a trusted foreman took prototypes to a handy field and there, in the light that would fall upon finished cars, finalised their shape by eye. What an eye!*

162 *Kimber's answer to the 2½-litre SS Jaguar of 1936 was the 2-litre, seen here in drophead coupé form, the bodybuilder being Tickford (Salmons, of Newport Pagnell, in fact). It was a very handsome car, with long, flowing wings, a long bonnet, within which there was a 6-cylinder Wolseley engine and a hood that half-opened, to* de ville *position, or could be folded down for open-air motoring.*

162a *The 1½-litre Aston-Martin had a most effective four-wheel, cable-operated braking system. The large-diameter drums were well-exposed to the air, as were the car's occupants. The 1933 Le Mans two- to four-seater, with its characteristic cycle-type wings (they move with the wheels), heavily-louvred bonnet and external exhaust down-pipes, is a fine specimen of the English sports-car.*

163 *The chassis of the Riley Nine was closer to the ground than was usual when it first appeared, which meant that the overall height of the car was kept down. The Monaco body was one of the first into which one stepped down. Deep footwells at the rear gave passengers plenty of leg room and the use of air cushions instead of the conventional stuffed and sprung seat cushion increased interior height without reducing comfort. The diamond motif seen here in window shapes and cut-outs on door cappings was as much a Riley feature at one time as the hexagon favoured by Kimber for MG cars.*

164 *The Riley Lincock was an attractive close-coupled saloon. By 1933, when this example was built, Riley front wings and running boards were united in smooth and flowing fashion and the radiator sloped outwards at its base. The 'eared' hub caps, of the 'knock-off' type which were loosened or tightened by clouting with a special hammer, the quick-release tank filler-cap and the spring-spoked steering wheel are visual reminders of the sporting nature of the car.*

shapes and proportions were made. apart from the Monaco [163], there were several other body types, mostly closed, of equal individuality [164]. That quality began to decline when Riley turned to steel bodywork in 1936, but even then their distinctive look was retained. It was lost, completely, when the company was swallowed up by Lord Nuffield (formerly Sir William Morris) in the late 'thirties' By some miracle there was a renaissance at the planning stage of the post-war Rileys and they matched the high visual qualities of their by-then distant forebears.

During the 'thirties, as the reader may have begun to guess, the appearance of the motor-car was being taken more seriously in Britain than ever before, following some distance behind the trend that had begun many years earlier in the United States and had spread to all other car-producing countries, affecting mechanical design, appearance and manufacturing method [165]. In Britain it had taken rather longer for any noticeable alterations to be made; the 'Bullnose' Morris, for example, changed little between 1919 and 1926; indeed, the main differences were the fitting of balloon tyres to the later cars and, in some cases, brakes to the front wheels. Apart from a change of radiator material and shape in 1916, and a certain softening of contours in the early 'twenties, the Ford 'T' looked much the same in 1927 as it had done eighteen years before. The flat-radiator successors to the 'Bullnose' lasted for five years, with only slight refinement of detail. The Model 'A' Ford—in production from 1927 until the Spring of 1932—filled-out in shape during that period [166] and, largely as the

165 *Even with simple jigs and a chain-operated track it was possible to turn out many cars in a working shift. In this pre-war view of a section of the Australian Holden factory two- and four-seat body shells are being prepared for painting, drawn to the spray booths on wheeled trolleys with tilting tops. The spaciousness of the section, and its tidiness, are unusual.*

166 *The Model 'A' Ford was a far better-looking car than any of its fifteen million predecessors, and this was an important factor in its sales success. The standard roadster, or two-seater, had good proportions and a neat appearance, and while it was in the contemporary fashion it retained a make identity that lasted for that decade.*

result of a well thought-out scheme of progressive development, became one of the best-looking mass-production cars of its day. Ford shapes were simple and not at all showy or pretentious; they were honest and straightforward and appropriately finished; the balance between plated surfaces and paintwork was nicely judged, and proportions overall were very good. Throughout the 'thirties the appearance of Ford motor-cars was always distinctive and far more restrained than that of most other American makes, even in the bad years, from 1934 onwards.

Before then, for six years or so, restraint had been a fairly general feature of the world's automobiles. Apart from the inevitable exceptions, good visual manners were widespread. There were plenty of exciting-looking cars or impressive-looking

167 *The Roesch Talbots, the ones made in London, that is to say, were fine-looking cars, whether bodied by their makers or by one or other of the outside specialists. The high-set, close-together, headlamps were characteristic of Talbot cars until the mid-thirties. This is an 18 hp model, its drophead body made, perhaps, by Ranalah.*

ones or downright handsome ones. Rolls-Royces, large or relatively small; Bentleys of all sizes; the $4\frac{1}{2}$-litre Invicta; almost any Sunbeam with its maker's coachwork; some Frazer Nashes; the 2-litre Lagonda; Talbots [167]; any Bugatti bodied with taste; the Alfa-Romeo 1,500 cc; the Mercedes-Benz 38/250 with open sports bodywork; Lincolns and some Packards. Other lists might be put up for reasons every bit as good.

Commercial vehicles and buses followed a similar pattern of design development. The gradual disappearance of the solid tyre and the spoked wheel and their replacement by the hefty-looking pneumatic and the steel-disc gave each kind of vehicle a more substantial look. A different type of chassis was coming into use for passenger-carrying; much lower than before, with an upsweep of the frame side-members over the rear axle to allow for spring and axle movement and, in the case of rear-entrance bodied buses, a long extension behind the back axle to support the platform. Fewer and fewer buses were of the bonneted, or normal-control pattern—except the smaller sort and some intended for touring, and forward-control, with the driver perched alongside the engine, was practically universal for service and long-distance vehicles. The number of bonneted lorries [168] was still quite high; for a time certain models featured a prominent bonnet that projected far ahead of the front axle, rather in the manner of some present-day earthmoving equipment. If this was to improve engine accessibility it was most successful, and Commers and Thornycrofts of that type are particularly remembered. Recognition of lorries and buses in those years was easy for the enthusiast large or small. Radiators stood out proudly and unmistakably; very large as a rule and almost always adorned with their maker's name and badge and some—Leyland particularly—carried a model name as well [169]. Personal favourites of the late 'twenties and most of the 'thirties were the Leylands, which were thought of as the Rolls-Royces of the lorry and bus worlds. Like Great Western Railway locomotives they carried names—names like Bull, Bison, Buffalo, Cub, Lion, Lioness, Tiger, Titan, even Hippo. Most unusually for their time, the lorries had a panel of grouped instruments, rather like those fitted to so

168 *Throughout the 'thirties the majority of lorries in the under 3-ton load class retained the traditional bonnet. One of the best-known was the Bedford, a General Motors product made in the Vauxhall plant at Luton. Its good appearance was no accident; its origins were American and there the look of a vehicle was taken as seriously as its mechanical specification.*

169 *By 1930 the earlier austerity of appearance of the average bus had gone, along with the solid tyre. Alexander's, of Falkirk, built this comely body on a Leyland Tiger chassis, and provided an opening roof, a large luggage space on the roof (reached by ladder) and alternate opening windows. Engine accessibility is good on one side only.*

many cars, and the standard of their cab amenities was above average. The Albion range was also favourably regarded, partly for personal reasons [170] partly because it was Scottish (and in the 'thirties one of the only two survivors of a once-large home motor industry), and partly because good appearance had been characteristic for some time. Albion buses, like Leylands, had names, too; Viking, Valiant [171], Venturer, Valkyrie. The AEC was distinctive; simple and honest, without affectation, in the Ford manner. There again model names were displayed—Matador, Regal, Reliant and Renown, for example. Thornycroft looked their best, perhaps, in the late 'twenties, when their shapes seemed to fit together in a most agreeable way; their wheels were interesting then, whether they were of the pierced-disc type, as on the J.2. lorry chassis, with solid tyres as surround, or of the later disc, with large domed hub-caps and what seemed like rows of retaining-nuts, and plump pneumatics. Their radiators are remembered as being very wide—perhaps to accommodate the name [172].

Mention has been made of the growth in demand for what used to be called the 'light commercial'. Between them Ford and Morris had a large share of the British market in the late 'twenties, and Chevrolet and some imports were also popular. A British-built challenger appeared in 1931, from the Vauxhall plant at Luton, and although many were heard to say that the new two-ton Bedford lacked stamina time proved them wrong, eventually. To some extent the new make had points in common with the American designed Chevrolet which had been built at Luton since 1929, but was, nonetheless, of excellent, distinctive lay-out, very good performance and low running costs. Properly cared-for, its life-span could be surprisingly long; one Scottish survivor, which is still in full working order, was in everyday use from 1932 until 1977, and other early models—buses as well as lorries—have become the treasured possessions of perceptive enthusiasts. A well-maintained two-ton truck is in its maker's collection at Luton and is not a static exhibit there but a vehicle taken out for an airing on occasions.

One remembers how quiet [175] those 6-cylinder Bedfords were in top gear; indeed they rivalled private cars in that respect and this was an achievement in the 'thirties. Until the coming of the diesel-engine lorry and bus, noise levels ranged from loud to very quiet—the trolley bus being by far the least obtrusive of the lot. A Scammell

170 *Road testing in fine weather can be exhilarating, but for the works driver it can have less pleasant aspects. The Albion rising sun symbol and distinctive nameplate top a tall, easily recognisable radiator. Behind the temporary front seat, large blocks of concrete are bolted to the chassis as a test load.*

171 *Its buses were almost as well-known at one time as the sturdy commercial vehicles made by Scotland's oldest surviving motor manufacturer. Among period details of this 1932 or '33 Valiant are decorative louvres, a sliding door, steps up the body side to the roof luggage container, an opening roof and head supports on each seat. The Albion name logo and rising sun symbol have been neatly combined in the radiator badge. The full-front is unusual for the period.*

172 *This six-wheeled Thornycroft is of the semi-forward control type, almost half of its bonnet protruding from the cab, along with part of the steering column. Features of interest are its tiny headlamps, the rivetted cab-front, the skylight and the neatly curved ladder to the tank top.*

Albion

170

WESTERN <SMT> Co. LTD

171

ESSO
BRITISH
OIL
DALKEITH
TEST WELL
ESSO
BRITISH
OIL
from the
DALKEITH
TEST WELL
THORNYCROFT
GG 4836

172

173 *Nine years after it was acquired in 1930, this Scammell articulated tanker was still in everyday use. Its makers were specialists in making articulated vehicles from the early 'twenties and have been responsible for some remarkable examples of the type. They favoured solid tyres for longer than most of their contemporaries.*

174 *A mid-thirties luxury coach; a Leyland Tiger built for long-distance touring and most lavishly furnished, therefore. The smooth, modern shape of the body is picked out in well-applied brush lining, the side windows open, there are ventilating louvres above each and the destination board is unobtrusively faired into the roof.*

articulated lorry [173], very heavily burdened, and crawling up a steep hill, made a fair commotion, but such sounds were much less disturbing, or unpleasing, than the din of a diesel pulling hard; much less difficult to tolerate on a long journey.

The writer is fully aware of the practical advantages of the oil engine; he is greatly attracted by the low fuel consumption and high life-expectation of certain types used in private cars. But he grew up during the period when petrol was still the prime moving spirit and when a long journey in a Leyland Tiger bus [174] was the nearest to Rolls-Royce travel a small boy in his circumstances might reasonably expect. He experienced the gradual replacement of the petrol engine in buses; an increase in noise and vibration levels; a decreased pleasure in bus travelling. Unlike Pope's child he was not 'Pleased with a rattle'. Such matters were of no interest to the operator, for whom the practical benefits of the oil engine, in terms of reduced fuel and maintenance costs, were of overwhelming importance. It was fortunate for those concerned that the introduction of the diesel [176] for road use coincided, more or less, with the start of the slump in the 'thirties. Certainly, the continuation of the latter greatly speeded up its development and widespread adoption by bus and lorry interests everywhere.

The history of the oil engine is too long to be dealt with in any detail here but one or two points of interest must be mentioned. The extent of British involvement in its design and evolution was very great indeed; the first-ever practical compression ignition engine was designed and built in 1890 by Hubert Akroyd Stuart and went into production, as the Hornsby-Akroyd, two years later—and five years before the first diesel engine was made. The builders were Richard Hornsby & Son. Although there was some attempt to produce engines of the diesel type in Switzerland and in Germany before the First World War, for road vehicle applications, it came to nothing. At the Berlin Show in 1924 the German firm of MAN, which had exhibited automotive diesels at the 1911 Turin show and at Berlin, in 1912, introduced a four-cylinder engine of about 45 hp that was quickly taken up. Other engines of the type were produced by Benz, in Germany (where the fact that petrol was in comparatively short supply stimulated interest in any alternative fuel) and by Saurer, in Switzerland.

175 *Not all main roads were as quiet in the mid-thirties. The place is Ware, in Hertfordshire, and this photograph was taken to show how unobtrusive the new concrete lighting columns were in a traditional setting. The lorry in the foreground is a Fordson; beyond it is a Bedford furniture van.*

176 *The Foden reputation was made originally with its excellent steam vehicles; in the early 'thirties, however, when the market for steam lorries was much reduced, the company began to make diesel-engined trucks. This one, built at the end of 1933, is still running. It is fitted with a Gardner engine. Its trailer is a Foden product; all in all, an impressive combination.*

Towards the end of the 'twenties tests began in Britain to determine the practical value and future potential of the oil engine for road vehicle use. Some experiments involved Gardner marine-diesels of modest power that seemed to possess the characteristics considered necessary—relatively light-weight, flexibility and a reasonable range of speed—and others were made with Benz-type engines of somewhat higher power. Their behaviour was carefully judged from bus and lorry installations and found so very promising that a period of intensive research and development followed.

By 1930 Gardner were busy building diesels of the highest quality, answering an already widespread demand, either as original equipment for new bus and lorry chassis or as replacements for existing petrol engines. At the same time, many individual manufacturers were developing their own oil engines, and by 1932 most British manufacturers listed these as options, a state of affairs that continued until the beginning of the war in 1939. The price difference was considerable: power for power a diesel cost about £400 more than a petrol engine in those days, but this was compensated by the very real saving in fuel consumption, simpler maintenance because of the elimination of the usual ignition and carburation systems and greatly extended intervals between overhauls.

Not all these advantages materialised at once; there were the inevitable teething troubles but the energy and inventive intelligence that was applied, especially in Britain, was immediate and effective. For the larger goods and passenger vehicles the practical advantages were irresistible, particularly from the cost angle. It was not simply that a well-maintained diesel used less fuel, but the fact that diesel oil was taxed at 1d per gallon, whereas the tax on the same quantity of petrol was 8d. Improvements in design and construction continued and it was found that the overhaul periods got farther and farther apart; cylinder heads stayed put for 100,000 or 200,000 miles, perhaps more; and it was often 300,000 miles before an engine strip-down. By petrol-engine standards, behaviour of that kind was quite unimaginable; lifting the head for decarbonising was a routine matter every 10,000 miles or so before the war, and it was often just as well to check the cylinder bores at that mileage. The owner of a small car, a Morris Minor or an Austin Seven, perhaps, boasted of the fact that he reached 15,000 miles before major attention to its little engine was necessary. It was not an expensive operation; £2 or £3 would cover the cost of boring the cylinders of a small 4-cylinder engine and fitting new pistons and rings.

Main-bearing renewal (the big-ends) was also a frequent necessity in the early 'thirties, though again it was comparatively inexpensive; and even as late as 1939 25,000 miles was considered to be a very satisfactory interval between overhauls. This was a variable, of course; the kind of use that a vehicle endured, and the manner in which it was treated, had a major influence on the way it performed and lasted. This was strikingly revealed by W. A. Robotham, then Chief Engineer of the Motor Car Division of Rolls-Royce Ltd, in a paper, 'Some Problems of Cylinder Bore Wear', which he read to the Institution of Automobile Engineers in January 1947. After suggesting that a reasonable universal target figure for rebore intervals should be 50,000 miles he detailed the kinds of difference that varying operational conditions could make. His company knew from its records that reboring might be necessary after as little as 15,000 miles of continuous London traffic driving. In what he called average English conditions, with fifty to one hundred-mile journeys and more frequent accelerating and decelerating, the rebore interval was more likely to be about 50,000 miles: whereas for cars driven at high speed on the Continent, it could extend to 100,000 miles. Robotham's averages were drawn from statistics covering a fifteen-year period from 1932. It would be extremely interesting to find comparable statistics for other makes of car of the time.

Technically speaking, progress was steady rather than spectacular during the 'thirties. Although basic engine design remained unaltered, continuing research and development raised power outputs without any accompanying increase either in noise or roughness; on the contrary, there was an all-round improvement in levels of quietness and smoothness during the decade. Mechanical efficiency was increased, and there the production of better fuels played a significant part. In this respect there had been useful and important lessons learned from theoretical and practical work with aero-engines and the very highly-developed engines of sports-and racing cars [177 and 178]. As far as car transmission systems were concerned, the most valuable single improvement was the general adoption of synchromesh, but another important development was the way in which gear-noise itself had been reduced. Although most motor manufacturers made their own gearboxes a certain number still used the services of such specialists as David Brown Ltd or Moss Gears.

177 *The first ERA (English Racing Automobiles) appeared in 1934, powerfully promoted by the late Raymond Mays who, along with many other British drivers, gained competition successes in succeeding years. The survival rate has been remarkable; no other make can excel it in that respect and it continues to compete successfully to the present day in historic racing-car events.*

178 *There was no ERA for road use but Aston-Martin, a make that also took part in racing in the 'thirties with its splendid 1½-litre, offered this Ulster model in 1935. It was a replica of the type that won the team prize in the 1934 Tourist Trophy race, and was sold with a guaranteed maximum of 100 mph. Its pointed tail houses the spare wheel and there is an impressive row of control switches along the instrument panel.*

Until the mid-'thirties Rolls-Royce made the greater part of its chassis—about the only components bought-in being sparking-plugs, batteries, tyres and instruments. After the death of Sir Henry Royce in 1933 there was a gradual relaxation and, eventually, Borg and Beck clutches, Marles steering gear and Stromberg carburetters featured in Rolls-Royce specifications. But the company continued to make most of its electrical equipment, and its own extremely good mechanical servo braking system. The drum-brake had been developed to a high degree of efficiency, to such an extent that very few cars of the period were faster than their brakes. One of the

179 *Cheap as it was, the Ford V-8 had great distinction of appearance. There was a good range of body styles on offer, including this rakish-looking roadster, and the performance by contemporary standards, was exciting. Note the semaphore-type direction indicators mounted on the screen pillars, the tiny sidelamps and the electric horns set below the headlamps. The girl is Miss Finland, 1934.*

exceptions was the Ford V-8 [179] in its first year or two of production, before something was done to improve a system adequate for the 65 mph model 'B' but scarcely powerful enough to manage it when enlivened by a potent 8-cylinder power-plant.

Some makers chose to use hydraulically-operated brakes, of a type that had been under intensive development since the early 'twenties (notably by the Lockheed company), while others preferred one or other of the mechanically-operated systems then available. The Bendix 'self-wrapping' brake was very widely used, both in America and in Europe. Its main competitor in Britain was the Girling system, an English design developed in the late 'twenties and used by a number of makes during the next decade. Instead of the cables commonly used to transmit driver-effort from the pedal to the brake-shoes it featured rods. It was very consistent in its behaviour, and highly efficient.

Once upon a time every road vehicle running on pneumatic tyres had as many Achilles heels as it had wheels. Its tyres were its largest single source of trouble and expense, as a rule, and a puncture- or trouble-free journey was something to brag about. By the 'thirties great improvements in tyre design, materials and construction had been made, which increased their strength and their resistance to wear. Punctures were still an inescapable fact of motoring life, but their frequency gradually declined [181]. There was another favourable factor, and that was the decreasing number of horse-drawn vehicles in daily use on British roads. In 1922 these totalled just under 233,000; by 1930 there were 52,400. As broken horse-shoes and their nails were the cause of many punctures, their reduction was good news for the motorist. Until the all-enveloping mudguard appeared, along with much lower bodywork, tyres and their treads were a quite conspicuous feature of the motor vehicle. In Britain by far the most familiar tyre was the square-block tread pattern favoured by Dunlop. Another distinctive type is remembered not so much for its tread as for the thin red line that ran round one side only. The owner who did not care for the colour simply arranged for the lined side to face inwards. India tyres of that kind were standard equipment on the new $3\frac{1}{2}$-litre Derby-built Bentley when it appeared in 1933 [182].

Wheels and tyres, it must be remembered, have always been important visual elements. The spoked, artillery-type wheel had been out of fashion for some time and in 1930 its only large-scale use was on the flat-radiator Morris Cowleys and Oxfords, and on Austin Twelve cars and taxis [183]. It had been replaced by the wire wheel and, to a lesser extent initially, by the steel disc [184 and 185]. One maker faithful to the disc from 1919 to 1939, apart from limited use of wire-wheels on certain models, was Armstrong-Siddeley. Another was Citroën [XXI]. It was a feature of the attractive little four-wheeled Morgan that was first seen in 1935. For goods and bus chassis there was no question of using anything else, and many handsome designs were produced.

In the mid-'thirties the pierced disc began a run of popularity that has not yet ended. It is easy to see why, as the type combines decorative and functional features

181 *The large Dunlop tyres and two-row spoked wheels, complete with balance weights, are an important visual element here. The car is a Rolls-Royce Phantom II Continental, circa 1932, with an exceptionally handsome sports saloon body by Park Ward. Boot space is restricted and, unusually, only one spare wheel is fitted to the special carrier at the rear.*

182 *The Rolls-Royce-built Bentley chassis was a gift to the enterprising coachbuilder because of its fine proportions, its low radiator, set between the front wheels and its long bonnet. Barker designed this beautiful coupé cabriolet de ville, shown here in the de ville position, that is to say with the fore part of the hood folded away into the box provided for its reception in the front of the main 'head', or hood. Note the three door hinges, the unobtrusive bumpers and the large boot. Because of its lack of structural rigidity it was very difficult to keep a body of this type rattle-free in extended use.*

183 *Tattered and torn, after at least fifteen years of arduous London service when it was photographed in 1949, but still thoroughly roadworthy; the mechanical stamina of the Austin Twelve taxi was matched by the durability of its bodywork, which came from a variety of sources. Its artillery wheels give it an attractive vintage look. This example has a landaulette body, with a luggage rack on the roof.*

184

185

186

at one and the same time. With an increasing amount of the car mechanism covered by bodywork reaching nearer and nearer the ground a proper circulation of air for cooling purposes was becoming more and more difficult to arrange [186]. The brake-drums were in direct need—the more so with the higher speed of the average car—and the piercing of wheel-discs was an effective way of directing cooling draughts in their general direction. Wire wheels looked very well, and in those days nothing else would have been considered for any self-respecting sports-car [187]—the 4/4 Morgan was an honourable exception. They were something of a nuisance to clean, however, and many touring-car owners covered them over with specially-made lightweight discs, which often improved the look of the car. Unfortunately, their presence discouraged regular examination of what lay behind, so that loose or broken spokes were often overlooked. Wheel covers, or 'spats', also hindered access [188].

184 *Steel disc-wheels were a feature of this rear-engined Mercedes chassis of the mid-thirties. It had a 'backbone' chassis, with outriggers to support the body, and independent suspension of all four wheels. The 4-cylinder, in-line engine was water-cooled.*

185 *Steel disc-wheels were a feature of this Straight-Eight Packard, along with prominent hub-caps and an easily-identified radiator, set well forward. The Barker body is a sedanca de ville, of semi-razor-edge type, photographed in de ville position, with the cover of the driving compartment stowed inside its box in the roof, above the speaking tube.*

186 *The forward-mounted radiator of the 12-cylinder Rolls-Royce Phantom III caused difficulties for the coachbuilders from 1936 onwards until the designers regained their visual balance, as it were. In 1937 Park Ward produced this sedanca de ville on chassis no. 3CP.192—an interesting body with oddly-shaped doors (that at the front has a pull-out handle), a metal cover above the front compartment instead of the more usual leather one, and close-fitting wings.*

187 *Wire wheels were still fitted to the new-look 'C' type Aston-Martin of 1939. The rest of this very fast car was vastly different in appearance from its predecessors, and did not please the traditionalists. The wire-mesh covered radiator opening and ventilating panel in the bonnet side have a makeshift and cost-saving look (above, left).*

188 *Even Italian designers made errors of aesthetic judgment at times. The appearance of this 1938 Alfa-Romeo fixed-head coupé, by Superleggera, is striking but one wonders if the covers of the rear wheels were left on when the car was racing. In place, they would certainly slow down wheel changes (above, right).*

More generally, the growing tendency to enclose made routine service or repairs more difficult to carry out. Oil- or grease-nipples were less accessible, for one thing, and for another the lower build of cars made it awkward to crawl underneath to take a look at the mechanics of the vehicle. Less and less was upkeep by the owner-driver encouraged: more and more was maintenance a task for the garage. The number of items that demanded regular attention was still high, as may be seen from examination of any pre-war instruction book. On some makes, Rolls-Royce, Bentley, Daimler and Rover among them, there was automatic chassis lubrication to minimise wear on the parts it supplied (with predetermined quantities of oil) and to reduce actual maintenance to the fewest possible operations [189 and 190]. On the whole, access to components requiring regular inspection and attention was good; the real troubles did not materialise until after the war, when there really was justification for complaint in many cases. It should be said that a properly cared-for vehicle of the 'thirties was unlikely to give its driver much trouble, either on or off the road; except in the earliest years, indeed, the same may be said with equal truth of all ages of the automobile.

Throughout the 'thirties the popularity of the 6-cylinder engine in Britain was high. Its former superiority in terms of inherent quietness and smoothness of

189 *It was not until the late 'thirties that the fashion for all-enveloping bodywork reached the higher grades of automobile. In 1934, when Barker offered this stylish foursome coupé limousine (a totally meaningless description, like so many invented by coachbuilders) much of the chassis was visible and accessible, which made maintenance schedules easy to carry out. With automatic chassis lubrication fitted (as it was to the 20/25 hp Rolls-Royce) these were less time-consuming than they had been in the 'twenties. Cellulose paint finishes made regular cleaning an easier operation than ever before. Although this body looks like a drophead its hood is fixed, for greater structural rigidity and longevity.*

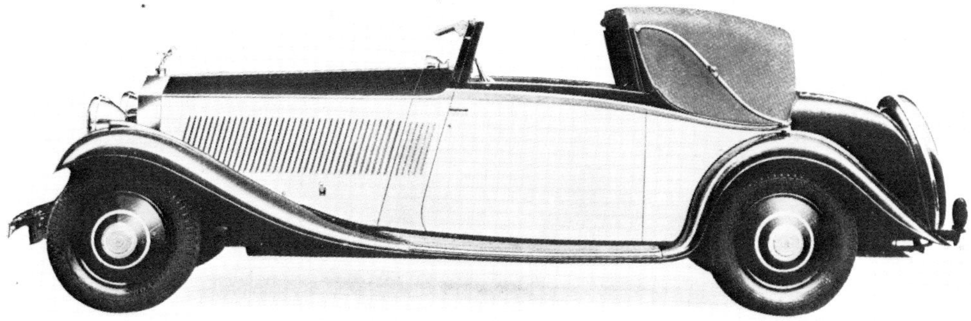

190 *Bodies of such elegance and such timeless design are a reminder of the very high design standards maintained by the best coachbuilders. Barker and Co were the creators of what they called a coupé cabriolet de ville, which is illustrated in its de ville position. Unlike its sister car in plate 189, it opens completely. The two-colour decorative scheme is well-suited to the overall shape of this magnificent vehicle.*

running was seriously challenged, however, as progress continued in the design of engines and their flexible-mountings and in techniques of sound insulation, and long before the end of the decade most 4-cylinder units were remarkably unobtrusive at low and medium speeds. The comfortable cruising speed of a car in the 8 to 12 hp range in the late 'twenties was around 40 mph, any increase leading to a pronounced increase in sound and 'palpable evidence of power' (or vibration, in a single word).

Ten years later the then-current Morris Eight cruised with ease, and quietly, at 50 mph, and its successor, the tank-like 'Series-E', would run up to 60 mph without mechanical stress. The same maker's sloping-back 10 and 12 hp models built between 1935 and 1938 were quite exceptionally quiet for their class, and on retesting after the war, when the cars in question were at least ten years old, they still impressed by their total lack of fuss. Other cars of the same type and period which proved good in this respect were the 'Flying' Standard 12 and 14, the integral-construction Vauxhall 10, the luxurious little Wolseley 10/40 and 12/48, the later Hillman Minx and its sister-in-disguise, the Sunbeam-Talbot Ten. Apart from the Seven which, until the end of its long production life, growled and occasionally tinkled, the other Austin models were singularly sweet-running. Indeed, their maker's advertising featured some of the methods of sound insulation then used. Perhaps the quietest small- and medium-power, medium-price cars of the period between 1934 and 1939 were the Rovers. They resulted from a crisis in the company's affairs which came to a head in the middle of the slump. The situation was taken in hand by the Wilks Brothers, with strong support from H. Howe Graham, a Birmingham accountant, and others, and a clearly stated policy giving first place to quality of design and manufacture was followed from 1932 onwards.

191 *Between 1933 and 1939 Rover cars were very good value for money in their class, their makers adhering to a policy of standardisation of engines, transmission, chassis and bodywork that held their production costs at a most reasonable level. During the brief spell of popularity of the so-called streamlined saloon, mainly in 1934 and 1935, this 'Speed Fourteen' streamline four-door coupé, to give it its catalogue description, was offered, at £415. It was on the standard 9 ft 7 in chassis but its engine was fitted with three carburettors and other modifications making it faster. Its headlamps and foglamp were grouped closely together, and it had a chromium-plated radiator stoneguard.*

Those pre-war Rovers were extremely good value for money [191]; they were handsome, fully-equipped, quiet and smooth—yet lively and fast—and, apart from their short-term economy as far as fuel consumption was concerned, they proved to be remarkably durable. After the war they were still fetching high prices second-hand, their mechanism and structure unaffected by time, it seemed. The writer's first experience of quietness in a motor car took place in 1938, when he was able to drive a 1936 Rover 14. Some impressions have not been forgotten: the comfort of the front seats, for example, and the way in which the outside world was shut off, as it were, once the doors were closed. Then there was the display of convex-glassed Jaeger instruments to admire, the large-diameter steering wheel, with its slender rim and spring spokes, the short, rigid gear-lever rising from the central tunnel and the comprehensive forward view, through an unusually deep windscreen [192]. In first and second speeds there was a restrained hum, but in third no gear noise whatsoever and practically no evidence of engine. Even at 50 mph in third the car ran with a lack of fuss new and exciting for someone interested in mechanical refinement—and so completely unused to it. In top gear the way the car could hold 60 mph without any kind of effort was a revelation to someone whose previous motoring at such desperate velocities had been associated with engine roar, frenzy and all-enveloping vibration.

The gearchange, too, was a different matter from that of the 1930 Morris Cowley and others. Although these were easy enough to master, the floppiness of the long, baton-like levers was disconcerting and the howl of gears once engaged, displeasing. The Rover gear-lever was short and firm and did not flop about; its movements from

192 *While the streamline saloons and coupés were going out of fashion and favour some designers and manufacturers were creating closed bodies of pillarless type; the Humber Vogue was one; the Talbot Ten another; neither was particularly costly. Because of its inherent flexibility the chassis of the 3½-litre Bentley was not, perhaps, the most suitable foundation for a type of construction lacking the usual reinforcements of door pillars, but Gurney Nutting (among others) had the courage to try. The result, in their case, was this most attractive body.*

position to position were short and crisply-made, and no howls ensued. The change itself had to be judged accurately—there was no synchromesh—but for the timid driver help was at hand; a free-wheel was part of the equipment, and with it engaged clutchless changes, up or down, could be made confidently and without clash.

The steering of the Rover was a little disappointing; it seemed springy and resistant to alterations of direction. Nearly thirty years later extended ownership of 60 and 100 models revealed a similar characteristic, as if Rover designers had assumed that their cars would seldom deviate from straight-line running. A vast improvement was made when the 2,000 model appeared in 1963.

A free-wheel was part of the specification of all Rover cars in the new range for 1934 and was to remain a standard feature of all subsequent production cars until 1957. No other manufacturer used it for so long a time, and it must be said that few others provided the essential accompanying features. Its use allowed the car to free-wheel—or coast—whenever the throttle was closed, and in such circumstances there was no engine braking available. For that reason alone very effective and reliable brakes had to be fitted. If the difference between 'fixed-drive' and 'free-wheeling' were to be minimised the engine itself had to be unobtrusive to a degree. Moreover, a high standard of roadholding and control was essential, a car running free being less manageable under certain conditions than one 'driving'.

In the early 'thirties the free-wheel had a run of popularity, Studebaker being almost certainly the first maker to take it up. Quite a number of manufacturers in Britain adopted it but none with the same comprehensive enthusiasm as the Rover Company. Certain of the larger Morris models in 1934 and 1935 had free-wheels

193 *By the very nature of the manufacturing the pressed-steel body was of smooth and simple shape, and easy to clean in consequence. The Morris Ten, of 1933, was typical of its type and time in that respect. What is untypical about this particular car is the fact that it was photographed in Germany (with another kind of dogcart), continental motoring in English family cars being rare in the 'thirties.*

and, to complicate matters, 'traffic' clutches as well. The latter were of the centrifugal type that disengages automatically when engine speed falls below a predetermined figure. Riley also adopted the free-wheel, and, more recently, Citroën have shown a liking for the system. Its use in the early 2cv model and variants to date tends to pass unnoticed in the general profusion of their mechanical features. During its brief spell of popularity in the 'thirties the free-wheel made driving much easier and probably gave the makers of brake-linings additional trade. To some extent careful manipulation of the system could improve fuel consumption—by ten per cent, at best. Its passing was inevitable once synchromesh came into general use, yet, surprisingly, Rover kept on with the system even after they, too, belatedly adopted that universal aid to easy changing. Since the end of the war, the free-wheel has sometimes been used in conjunction with first gears lacking synchromesh, or in two-strokes such as the German DKW and the Saab.

The cars of the 'thirties were easier to drive than their predecessors; in some ways they were easier to maintain; they were easier to clean and, in spite of the greatly extended use of all-steel bodies by the largest producers, they seemed to last as long [193]. The writer remembers far more disreputable fabric bodies then; false-Weymanns that began to peel within a year or two and revealed their wooden skeletons. Rust occurred, inevitably, but it is not remembered as a conspicuous feature of the pre-war mass-produced, pressed-steel saloon. One factor that has to be taken into account is the fact that the majority of cars in those days spent their nights under cover, either in their owners' own garages or in public ones; indeed it was not until the 'fifties that it became legal to leave a car outside overnight, without lights.

The British market was dominated by the English car. Although certain Scottish manufacturers were nominally in business at the start of the decade they had turned out no cars for some time. Only Albion and Halley, the bus and lorry builders, were still active, and the life of the latter was shortly to end. The Argyll, which had been revived after the Armistice, had faded away by 1928, its post-war productions unexciting, and Arrol-Johnston, with a 15.9 hp chassis that had far too much in common with its pre-war forebear, had no chance of survival. Beardmore taxis kept one Scottish name before the public for many more years, however.

The contraction of the industry that had begun in earnest during the 1921 slump continued. Without any doubt the most serious, most lamented casualty was the Bentley company. Although it had been under-capitalised since its beginnings, in 1919, its founder, W. O. Bentley, and an unusually dedicated staff had kept its products and their performance, on road and track, constantly in the public eye. Deliveries (of the 3-litre model) began in 1921; ten years later, when the Receiver was called in, the still-small concern was building a very large 8-litre car, which had been selling well, even at the height of the second slump since the Armistice, and another with the same chassis, under-engined with a 4-litre unit, which did not.

Perhaps it was an act of irresponsible folly to introduce another high-power high-cost car at such a time [194]. When the biggest Bentley appeared even Daimler had been compelled to cut its range drastically to seven models for the 1931 season. Their largest chassis was then the 13 ft 1 in wheelbase Double-Six, which had a 6.5-litre, V-12, sleeve-valve engine, and cost £1,550, bodywork adding anything from £750 to over £1,000, according to type. The Lanchester Straight-8 had been selling quite well, at a chassis price of £1,325. It is permissible, perhaps, to include the 3-litre Lagonda (in long chassis form, at £745, it was one of the costliest cars of its kind), and the similarly priced 25 hp Sunbeam.

From the Continent there was some strong competition. The superb 45 hp, 7,983 cc Hispano-Suiza cost £1,950 whether one chose the short- or long-wheelbase model (11 ft 1 in or 12 ft 1 in). The big 7,372 cc, Straight-8 Isotta-Fraschini chassis was priced at £1,750, which seemed almost reasonable when compared with the £2,185 demanded for the 45/280 Mercedes-Benz, a most substantial machine. A year later there was an even larger car on offer—the 62.96 hp, 12-cylinder Maybach 'Zeppelin'.

The American invasion of the British luxury car market was probably small, numerically speaking, but the quality of her top cars was high, and that included their coachwork. In 1931 the V-16 Cadillac, of 7,406 cc, was rated at 57.5 hp for taxation purposes here (its claimed output was 185 bhp), and its chassis cost £1,500. The Ford-owned Lincoln, with a 6.3-litre 8-cylinder engine, was available in two chassis lengths, 11 ft 4 in and 12 ft 6 in, at £980 and £1,050 respectively. Finally the Packard 39.4 hp, 6,318 cc, 8-cylinder chassis, long or short, was excellent value at either £1,195 or £1,050. Restraint, or visual good manners, was

194 *The 8-litre Bentley was one of the most impressive-looking cars ever built; its tall radiator flanked by Lucas P-100 headlamps and its long bonnet sides well-louvred, to assist in the cooling of such an enormous engine. The body of this car is a Freestone & Webb limousine. There was no weight limit on the 8-litre chassis; with more than 200 bhp available, it was unnecessary in practice (above).*

194a *The convertible was a body type always treated sympathetically by American designers and bodybuilders, regardless, almost, of price. This 1938 Lincoln V-12 has a generally elegant 'Convertible Victoria' by Brunn. The 'side-mount' spare-wheel and white tyre sidewalls are typical.*

194b *The 1937 Lincoln has a roadster body by Le Baron, simple in shape and easy to maintain, therefore. With the hood up, visibility is restricted.*

195 *The largest Bentley was built to compete head-on with the Phantom II Continental, the Rolls-Royce specially designed for high-speed foreign touring. The prototype is still on the road, in superb condition and one of the finest-looking cars ever built. The body was designed by H. I. F. Evernden, in association with Sir Henry Royce and under the inspiration, to some small extent, of the Riley Nine, and was built by Barker.*

characteristic of most of the largest, most costly American cars throughout this decade—in particular those fitted with custom coachwork—as it was of comparable European cars. Plates 194a and 194b show good examples.

As far as home competition to the new 8-litre Bentley was concerned, its main rival was the current Phantom II Rolls-Royce [195] which cost £1,850 in short, 12 ft wheelbase form. That, as it happened, was the same price as the big Bentley, in either 12 or 13 ft wheelbase lengths. The engine of the Rolls-Royce was an overhead-valve six, with crossflow head, that was designed for quietness and smoothness and the production of a reasonable amount of power—somewhere in the region of 140 bhp, it is believed. It had a very fine top-gear performance, a much more useful

194a

194b

195

196 *The sheer size of this 1933 Phantom II Rolls-Royce is impressive. Its shape is striking also, even though it is concealed to some degree by the flamboyant restoration in gold metallic paint. The white sidewall tyres suggest transatlantic ownership. But the car is original enough; a 1933 Thrupp and Maberly drophead with period Grebel spotlight and ship-type scuttle ventilators, side-mounted spare wheels, a warning siren between the auxiliary lamps below the big P100 headlamps and an interesting registration number—ALF222.*

set of intermediates than its predecessors and a much easier gearchange. It was also a simple car to look after, automatic chassis lubrication being one of its new features (in passing it should be said that this had been standard equipment on the American Rolls-Royce for some years). It provided an almost unmatched platform for some of the best-looking coachwork of its time [196].

The 8-litre Bentley was also propelled by a 6-cylinder engine but its designer had concentrated on quietness and smoothness and the production of a great deal of power, something of the order of 220 bhp being claimed, at 3,500 rpm. Its overhead valves were driven by an overhead camshaft which, in turn, was driven by three connecting-rods that linked little three-throw crankshafts at the end of the engine crankshaft itself and at the rear end of the camshaft. It was a rather complicated system, but it was efficient and very quiet—and that, after all, is what Bentley wanted. The result was a machine that could spin along on part-throttle at 80 mph and go on to exceed 100 mph—a performance that, in combination with great accelerative powers, made it quite certainly the fastest luxury car built in Britain. The Phantom II had a top speed of between 85 and 95 mph according to the length of its wheelbase, the type of bodywork fitted and, one suspects, the standing of its owner at Derby. Some of the short-chassis cars were very fast.

Although the writer has owned a Phantom II and driven many others, of various ages, he has no road experience whatsoever of the 8-litre. He has, however, driven a good 'Speed Six' Bentley, which he found to be very quick and very quiet, and if the larger car was even better in these respects it must have been exceptional. It has been suggested that it was superior to the Rolls-Royce; quite possibly it was. The 8-litre had a brief life, and only one hundred chassis were built; sixty-three of those between October 1930, when it was displayed for the first time at the Olympia Motor Show, in London, and the following summer, and the remaining thirty-seven at a later date. The time was wrong: Bentley Motors Ltd had to go into liquidation and one of the most interesting, and certainly the best of the post-Armistice newcomers was lost to the British motor industry. But not for long. The name was revived in 1933, two years after Rolls-Royce Ltd acquired the company from the Receiver, and although the new 3½-litre Bentley had much in common with the current 20/25 hp Rolls-Royce model, its performance was of a far more spirited and lively kind. Bentley competition successes at Le Mans and elsewhere during the 'twenties and into the 'thirties, had done a great deal for Britain's international reputation at a time when her involvement in motor racing on a wider scale was not at all extensive. The Bentley achievement then confirmed the great value of a development programme which had made proper use of information gained in racing to improve the standard product [197]. Bentley himself, always maintained that his company's racing activities were not expensive. Perhaps someone, some day, will produce precise costings. They will make interesting reading.

The important part that Bentley successes had played in publicising British cars was continued by another small company, with smaller cars of an entirely different type. The first MG Midget was a slightly modified Morris Minor fitted with an attractive, open two-seater, fabric-covered body that weighed, and cost, very little. Its 847 cc single overhead-camshaft engine produced 20 bhp, which was quite enough to give so light a car a brisk performance by contemporary standards. Its potential was greater, however, and almost the first thing that H. N. Charles did after

197 *The design life of the low-sided touring-car type of body was a long one—for proof compare the Austin shown in chapter 2 [62] with the supercharged 4½-litre Bentley illustrated here. Their basic shapes are similar. The 1929–30 car was developed by its former owner, the late Harry Rose, to a high degree of mechanical efficiency, but unlike some enthusiasts for speed he took equal trouble with the body work. It is not simply a 'clean machine'; it is a gleaming one.*

joining the MG Car Company as chief designer was to change the camshaft, which added 7 bhp.

This was only the beginning; there was far more to come, a fact quickly proved by an exceptionally able design team. On the Brooklands race track, in reliability trials, in the Monte Carlo Rally, at Le Mans, in hill-climbs and, eventually, in record-breaking, the little cars proved repeatedly successful. Again, it was a matter of competition participation improving the product. Wisely, however, the state of tune of the production 'M' type was maintained at a mild level. For everyday use it was untemperamental, reliable (dynamo-drive excepted), economical and enjoyable. For the serious and ambitious driver, more highly-tuned engines were available, in modified chassis, and this practice was followed in principle until well after the Second World War. The scope for development was great; the supercharged engine of the costly 'Montlhéry' model (which was of 746 cc, to qualify it for competition in the 750 cc class) produced 60 bhp, and even more power was obtained from the engines specially created for record-breaking.

The original Midget was a likeable little car that became immediately popular, offering strong competition to the Austin Seven Sports which, at £185 unblown, was £10 dearer. The two makes were to be up against one another for some years but the Austin, as a make, was never as widely-known for its sporting models. How much the racing rivalry was due to the known lack of accord between Morris and Herbert Austin is a matter for speculation. To some extent this personal animosity stemmed from the fact that Morris succeeded in buying the bankrupt Wolseley Motors for £730,000, after the Receiver had been called in towards the end of 1926. Austin wanted Wolseley, perhaps for sentimental reasons in view of his early association with the make; Morris was afraid that the concern might pass into American ownership, or so it has been said. The Vauxhall company had run into financial trouble a year or so earlier, and it would be interesting to know if he made any attempt to keep the firm out of the hands of the eventual purchaser, General Motors Ltd.

This was the first direct foreign move into the British motor industry, on a wholly-owned basis. The Pressed Steel Company was a joint Anglo-American enterprise. After the death of the Hon. C. S. Rolls in an aeroplane crash in 1910, the Canadian, Sir Max Aitken (later Lord Beaverbrook), acquired an important block of Rolls-Royce shares and soon made himself unpopular with the Board by suggesting that the company should go in for mass-production. At a later date he parted with his interest to the American tobacco tycoon, J. B. Duke, who returned them (profitably, no doubt) to United Kingdom ownership, some time after the war. Although Rolls-Royce connections with the United States have been numerous and close for a great many years, and American sales now account for one-third of present-day production, the company is very much an all-British one.

Apart from Vauxhall/Bedford/General Motors, the other major American car manufacturer with a large-scale investment in Britain is the Ford Motor Company, whose long-standing involvement with the British motor industry began in 1911, with the opening of its first assembly-plant at Trafford Park, Manchester. During the First World War, parts manufacture began there, partly because of new import duties and partly because of the great risk to consignments shipped across from the States. In 1919 production began at a newly-opened assembly-plant in Cork, Southern Ireland, and there too, eventually, component and engine manufacture were carried out. Trafford Park was closed down in 1932 and since then, a major part of Ford's United Kingdom activities has been centred on a vast complex at Dagenham, in Essex.

Changes in the structure of the car taxation system that became effective in 1921 had an adverse effect on Model 'T' sales in this country. Previously the car, which was rated at 22.4 hp, had been taxed at £6 6s 0d per annum, but under the new rules owners had to pay £23, instead. At once the makers of British cars of lesser rated power found themselves with a powerful sales asset, the Austin Twelve, for example, carrying an annual burden of £13, the Morris Cowley £12 and the Oxford £14.

The Ford Ton Truck was not affected in the same way because goods vehicles were taxed according to weight, unladen, on a scale that ranged from £10 to £30. Even after the introduction of the new Model 'A', in 1927, British sales were low; Morris, Austin, Singer and Standard had the major share of the family-car market, and, in the circumstances it was hardly surprising that Ford only managed to sell 6,224 cars and lorries here in 1928. The answer was not the hasty introduction of a 14.9 hp option, made in the Cork plant; the position grew worse and by October 1931 Ford was close to financial failure in England.

The company was saved by the rapid development of a small car at the Ford Dearborn plant, the first model that Ford had ever created specifically for a foreign market with its own special requirements. The Model 'Y' [198], was a high-set and

198 *It was not only the public that took the perky little Ford seriously; its introduction in 1932 was taken very seriously indeed by Austin and Morris, then the largest producers of small cars in Britain. Morris answered the threat from Dagenham by introducing his own Eight in late 1934, and a year or two later Singer brought out a Nine that, like the Morris, bore more than a passing resemblance to the Ford.*

amply-curved little 8 hp car accurately aimed at the still-growing market for small, economical vehicles and, at £120, it was cheaper than its main competitors, the Austin Seven and the Morris Minor. It was roomier, much livelier and probably more economical in the long term. It was easy to drive and at a great advantage in the faster-moving traffic of the day because of its acceleration, which whisked it up to 40 mph with a characteristic roar. Apart from its two-point suspension, which was of 1909 design (and made it a handy car for off-road motoring), it was a more up-to-date vehicle than its rivals, in many respects.

Fuel consumption was anywhere between 35 and 40 mpg and, because it was a Ford, parts and servicing cost very little; to read the published schedules is a nostalgic business. Alexanders, the principal Edinburgh agents, had an imaginative service before the war. Their customer was invited to leave his 8 with their workshops at tea-time and go off for a meal and an evening in the cinema. When he collected his car two or three hours later its engine had been removed and replaced by a reconditioned unit, for a matter of £6 or so [199].

The 'Y' sold well and for a time there was a Popular version that cost only £100. Two years after production began at Dagenham, in September 1932, the small, boxy and bouncy Morris Minor was superseded by a new Eight, the look of which was not unlike that of the Ford 8. Its appearance was more appealing, it was supported at eight points, by semi-elliptic springs, it had hydraulic brakes a good deal better than the mechanically operated ones of the Ford, and a sweeter, quieter engine. It was a great success.

199 *The Rolls-Royce 12-cylinder Phantom III chassis cost £1,850 and bodywork of this type (an enclosed limousine by Barker, to the order of HRH The Duke of Kent, in 1936) could cost £1,000 or more. Each of its kind was first-class value for money. The Barker body illustrated had many special features—an interior division, for example, locks on each door and on the petrol tank filler cap, three hinges on each rear door and concealed hinges on the front doors, radio and chromium-plated instrument panel.*

200 *Falling sales startled the Austin Motor company into action during the second half of the 'thirties. Appearance design, which had been neglected for some years, was taken seriously and the new cars that began to appear in 1936, with their pierced disc-wheels and large-section tyres, their smoother looking body shapes and their forward-mounted radiators, sold well. The well-equipped Ten, with sunshine roof and safety glass all round, cost a mere £178.*

201 *Austin 10 body assembly in Australia, on a moving line, or track, in the 'thirties. In the absence of positive information with the photograph it has to be assumed that the parts were sent out from England in the first place for local assembly, thereby receiving favourable import duty consideration.*

Austin became self-conscious about appearance in the mid-'thirties, fruitlessly at first. They tried again, with better results, in 1936 when a new range of sports-saloons [200] with a more positive and up-to-date shape than their archaic-looking predecessors [201] was introduced. Only the largest and smallest models remained unsatisfactory—and to make anything of the Seven, as it then was, would have baffled Pininfarina.

By 1938 the Ford 8 was greatly changed in shape [202], with a swept-tail saloon body of rather restrained form and front wings that were boldly curved, and pressed-steel, pierced disc-wheels instead of the wire wheels formerly fitted. Underneath it was still much the same as before—simply, sturdy, workmanlike. Another rival appeared in that year—the Standard 8, which is remembered as a very quiet, spririted little car, more comfortable than its class contemporaries. It had independent front-wheel suspension (i.f.s.), which gave it the distinction of being the first British car of its power to be so equipped.

I.f.s. was not new of course; indeed, its origins were Victorian, and French, but the best-known early English example was the type of sliding-pillar system used on Morgan three-wheelers from 1910, or thereabouts, and, suitably modified, by Lancia, on their Lambda models in the 'twenties. It made an impact in the States in 1934, when several makers announced new cars with what became generally known as 'knee-action' front suspension. Gradually, it was adopted by the majority of American manufacturers (Ford was one of the minority), and was also taken up by a small number of British and other European manufacturers. I.f.s. certainly improved ride comfort in most cases but because it allowed softer, which is to say more flexible, front springs there was a tendency for increased roll on corners taken quickly.

Alvis adopted i.f.s. in 1934, for their Speed Twenty model and used it for all their later, larger cars [203]. The unsuccessful Sunbeam Dawn had a somewhat similar system, using a single transverse spring. Singer tried out the Armstrong design but did not persevere; Vauxhall, on the other hand, used the French Dubonnet system in which the hubs were securely fixed to the front ends of fore-and-aft suspension arms, these in turn being pivoted for vertical wheel movement in trailing members that incorporated a torsional springing medium. Because the trailing members were carried on king-pins at the ends of a chassis cross-member, the entire suspension system turned with the steering. This is seen on its 12 and 14 hp models from 1935 [204], and on the integral-construction 10 from 1937. Rolls-Royce used a highly refined General Motors double-wishbone type for their splendid V-12 Phantom III,

202 *In* 1938 *the Ford 8 was a much-changed motor-car externally, with a swept back, an entirely new radiator and pierced disc-wheels. With few changes it remained in production until the 'fifties, its high and narrow body taking a surprisingly long time to look out of date.*

203 *Although Alvis began to use i.f.s. as early as* 1934 *they kept their radiators where they belonged—in line with the front-wheel centres—and retained the classic look. This model (probably a* 1937 *Speed* 25*) is low in built but there is unusual headroom inside its swept-tail, four-light body (above).*

204 *Not all drophead coupés were built on costly chassis; Vauxhall collaborated with the Grosvenor Carriage Co in the production of this pleasantly-proportioned body on their independently sprung* 14 *hp chassis. It is a* 1937 *model.*

from its introduction in 1935, and for the smaller Wraith from 1938 to 1939. With some changes the same system reappeared on the post-war Silver Wraith and the Mark VI Bentley. It was a feature of the few Mark V Bentleys produced in 1939.

Lagonda was edging up-market in the mid-'thirties, once W. O. Bentley's service contract with Rolls-Royce ended, and he went to Staines with ambitions. Apart from soothing and smoothening the current and quite conventional 4½-litre model, which was in direct competition with the Rolls-Royce-built 3½- and 4¼-Bentleys, he had plans for a 12-cylinder car to compete with the 'Phantom III'. He wasted no time at all, and his 42 hp V-12 [205] was the Show sensation of 1937. One of its most interesting chassis features was i.f.s, by long, fore-and-aft torsion bars, and contemporary reports all praise the exceptional combination of comfort and roadholding that it provided. In that respect the new V-12 and its sister 6-cylinder model were significantly ahead of the current Bentley 4¼-litre, which had conventional suspension.

Those Lagondas were supplied with first-class bodywork, of their maker's design and construction, and modestly priced. For example, the short-chassis V-12 cost £1,200, in 1937. The standard saloon [XXII] was listed at £1,550 and the equally distinguished drophead-coupé at £1,575. For comparison, the 4¼ Bentley chassis cost £1,150 and the complete standard-steel sports-saloon by Park Ward brought that price up to £1,510. Of the two, the Lagonda was the better-looking.

205 *Standard Lagonda bodies, especially after W. O. Bentley joined the firm from Rolls-Royce, in 1935, were good-looking and very well-equipped. Best of all were those designed for the new Bentley-designed 6- and 12-cylinder cars of 1938 and 1939. Here is the 12-cylinder drophead, the curves of its wings assertive but aesthetically satisfying.*

206 *During the second half of the 'thirties designers either concentrated on curves (and when they did so properly, as in the case of the Lagondas of 1938 and 1939, all was well) or the less popular razor-edge style. At the time a few Horch cars came into Britain from Germany and this drophead, with its strongly-emphasised hood and hood irons, is typical of German practice then. It is a type of body that has always been especially well-designed and made in that country.*

207 *When George Brough decided to make cars as well as his superb motorcycles he took especial care with their bodywork, designing it for long and troublefree life. Acherley, of Birmingham, were the makers of this drophead, which has a rather heavy look towards the rear. The Brough-Superior car was a mixture of English and American ideas and parts; mostly its mechanism was American; it was large by British standards and always lightly stressed, therefore.*

From 1934 onwards there was a gradual increase in the use of curves in body design, which was to lead to a steady decrease in aesthetic standards. The designer, who had been disciplined previously by the limitations of the pressed-steel process of manufacture, now found that there was very little that he could not demand of the die-makers; almost any shape, or combination of shapes, that he required could be realised. The result was an aesthetic catastrophe, in many cases. In the lower price range of family-cars design was better in Britain than in many other countries at the time and at the top end of the market those involved avoided the worst excesses, but in between, in terms of price, design quality was more uneven and often rather heavy [206 and 207]. The designers who had taken a second look at the achievements of

208 *Perhaps some designers adopted the razor-edge style in protest against excessive use of curves by so many of their contemporaries. Perhaps, on the other hand, they had been taking a second look at the way in which straight lines and subtle curves had been used with such effect by the eighteenth, and nineteenth-century coachbuilders. The 4¼-litre Bentley body is not altogether successful overall though its presence is considerable.*

209 *The Freestone & Webb fixed-head coupé on a 25/30 hp Rolls-Royce chassis is a better design—but visibility to the rear is hardly a good feature. If the luggage rack at the rear is any guide boot space must be limited. So, too, is room in the rear compartment.*

their forebears in the eighteenth and nineteenth centuries and evolved razor-edge styling to suit the automobile probably produced the most interesting work [208 and 209], showing examples of above average quality.

Britain avoided the worst excesses. America did not; she created them. It was a bad period, but it was inevitable; man had been given new tools to play with, once the techniques of all-steel bodybuilding had been satisfactorily evolved and twenty years later he was still too undisciplined in their use. In time the matter resolved itself.

6 *The Second World War* 1940 *to* 1945

When war was declared in September 1939 several British manufacturers had already announced new cars for 1940 that differed little from their current models. The uncertainties of the time make this scarcely surprising; in the mid-'thirties the possibility of a future war had been taken so seriously by the Government of the day that it had invited a number of important firms to take part in its Shadow Scheme. This was a plan to expand British aeroplane and aero-engine production by creating a number of 'Shadow Factories' that were to be designed, built, equipped, staffed and managed by the companies concerned (Austin, Daimler, Rootes, Rover and Standard among them), and run as separate entities. The Government was to be responsible for their cost. Many—but not all—were sited well away from built-up areas or existing industrial sites.

There was no delay in implementing this imaginative scheme—the first Rover Shadow Factory, for example, was turning out Bristol 'Hercules' aero-engine parts by the summer of 1937, and the second one was in production towards the end of 1940. By then the Rover company, which had stopped making cars at the end of 1939, was mainly concerned with aero-engines and parts, and with air-frames. In spite of serious fire and bomb damage to Rover's plants at Coventry and Solihull, output was increasing, but as a precaution there was a planned dispersal of production to an 'underground factory somewhere in the Midlands' and to a number of cotton-mills in the North of England. As the war continued the workforce grew considerably; the company was running six Shadow Factories and had almost a score of other plants under its direction. The part played by Rover in the scheme was fairly typical. However, its high reputation for quality involved it in something rather special; between 1940 and 1942 Rover was very closely concerned with jet engine development, and although this side of its activities was transferred to Rolls-Royce early in 1943, as a matter of official policy, an interest in the matter was retained.

After initial contact between Maurice Wilks, of Rover, and the inventor of the jet, Wing Commander Whittle, in 1940, the company received a contract for the production of development engines, for investigation of methods of production and for analysis of the two design types of the first development units. It was an enormously difficult task because so little was known about the subject; precedents were very few indeed, and to a great extent research and production were one and

210 *The pre-war Vauxhall 10 was a lightweight family-saloon with integral construction, a good performance and excellent fuel economy. During the war its makers turned out another integral-construction vehicle—the 'Churchill' tank—which was a heavyweight, a mobile weapon with fine cross-country performance and excellent fighting capabilities, but not at all economical.*

211 *The Bedford QL three-tonner was enormously versatile and adaptable. Its metal body was extremely tough and resistant to the wear and tear of war, its powerful, reliable engine and four-wheel drive could take it over difficult ground with ease and it was possible to tackle rough going with confidence because of the high ground clearance.*

the same process. There was a close collaboration with Joseph Lucas Ltd, another long-established concern, whose connections with the motor industry have always been highly important, and both firms were to be associated with gas turbine development for peaceful purposes once the war was over.

By the end of 1942 about twenty finished engines were on test; it was then that the authorities decided that this work should be handled by an aero-engine specialist. By friendly agreement Rover handed over their jet engine work to Rolls-Royce, and in exchange they were invited to take over that company's 'Meteor' tank engine project. Perhaps this V-12 was simpler to tackle than the gas turbine but it was by no means as straightforward as the 4- and 6-cylinder petrol engines to which Rover were accustomed. It was used in the 'Centurion' tank at first, and a later model with petrol-injection was used in the 'Conqueror'. A V-8 version, the 'Meteorite', was produced in petrol and diesel forms, and was fitted to the Thornycroft 'Antar' tank transporter and to a number of experimental vehicles, while special marine types were installed in sea-going vessels of various kinds.

Tanks were very much the concern of Vauxhall Motors, at Luton, along with trucks. In 1940 the company was asked to design and make a new tank, and within a year the 'Churchill' [210], a thirty-eight ton mammoth, was clanking off the production line. With ten other factories to help, close on 6,000 of these effective fighting vehicles were built. Even more impressive, in terms of numbers, was the total of trucks produced—almost 250,000 of a great variety of types, the most celebrated of which was the four-wheel drive QL [211] a square-fronted model that was to become a familiar sight in post-war years in civilian hands. Another well-known Bedford was the articulated 'Queen Mary' [212] that the Royal Air Force favoured for so many of its transport tasks.

212 *One of the most useful, successful and familiar wartime vehicles was the specially-designed 'Queen Mary' aircraft trailer, seen here under tow by a Bedford OXC tractor. Because so much thought had gone into its design it was enormously versatile in service.*

213 *Although the Austin ambulance was not the only one produced here it is the one best remembered. This well-kept survivor (now cherished by a member of the growing number of military vehicle enthusiasts) shows the clever design of its body—largely built-up with flat panels and, therefore, quickly assembled.*

For a time, Vauxhall was also closely involved with jet-engine work and then with much other aircraft activity, weapons and radar; in between times presumably about one hundred of the integral construction 10 hp cars were built. Bedford military vehicles are still in production and are used by many different military forces.

Production of SS cars continued until 1940 but their maker's increasing concern was with trailer manufacture and then with aircraft work of different kinds. Nearby, in Coventry, Standard were also engaged in making airframes and 'the Daimler', which had carried out such a variety of work during the First World War, was again heavily involved. This time it was responsible for the technically intriguing four-wheel drive, all-independent suspension Scout car and a larger armoured car, both of which could cover rough terrain with impressive ease. The latter was fitted with disc-brakes. An unexpected refinement was the use of the fluid flywheel and the well-known 'self-changing' epicyclic gearbox which had been a feature, also, of Daimler buses. Of the latter, incidentally, over 1,000 were made during the war, Daimler, along with Bedford and Guy, being allowed by the Government to build vehicles to Ministry of Supply specifications. Those wartime buses are remembered for their simple, but pleasing shape and for the hardness of their wood-slatted seats.

Memorable, also, are the really large tank transporters turned out by Albion and Leyland and Thornycroft, for example; the big AEC gun-tractors, the Bren carriers that were so plentiful at the beginning of the war, powered in the main by the

214 *This undated engraving shows what may well have been a mobile mediaeval military oven, massively constructed and visually interesting. At the beginning of the twentieth century, the Birmingham firm of Fisher & Ludlow already had a military connection; it was making military mess-tins in large numbers for use in the Boer War. During the First and Second World Wars it was also busily engaged in producing warlike stores.*

unobtrusive Ford V-8 engine, and the American trucks that were driven so quietly by the same means; the Ford and Willys 'Jeeps' that flitted about, top-heavy with troops; the handsome six-wheeled Crossley fire engines used by the RAF; Morris lorries and the pre-war 12 hp car in khaki; the neat little Austin Eight tourer, so popular with Army officers, and the Austin trucks and ambulances [213]; Hillman Minxes, Humber Super Snipes on outsize cross-country tyres (that created a quite distinctive whine on smooth-surfaced roads) and a distinctive four-wheel drive all-purpose vehicle from the same maker; Standard 12 hp tenders and many, many more. One wonders how many users of these vehicles had any idea at all that twenty years and more after the war was over, examples would be collected by military enthusiasts [214].

In spite of the enormous pressures of wartime work, time was found for consideration of what might be manufactured once the fighting had ended. No one could foresee the physical, let alone economic conditions of the warring nations, one peace returned, but the cumulative effect of years of inadequate road-maintenance would have to be dealt with. Good suspension was going to be one priority. Another was going to be even wider adoption of integral construction, its practical advantages having been effectively demonstrated by Citroën, Morris and Vauxhall although in small quantities. Generally speaking, integral construction was more suitable for Europe than for America, largely because of the different marketing policies practised across the Atlantic; with the traditional chassis as base it was possible to make frequent model changes in an economic way.

And yet, when entirely new British cars appeared from 1945 onwards they were seen to have separate frames, a fact which can be readily explained; each was of the fairly expensive, low-production kind. It was a year or two before the largest producers—Austin, Morris, Rootes, Standard and Vauxhall—began to announce their first new post-war models, and the majority were of unitary build.

As far as the interested motorist was concerned, in particular the reader of the motoring journals, much of what was presented in *The Autocar*, *The Motor* and *Motor Sport* referred to the past; car-making in Britain and elsewhere had stopped, for all practical purposes; there was a limit to the amount of speculative writing about post-war possibilities and, as a result, well-informed writers had a field-day. Undoubtedly this had a strong influence on many readers, some of whom were traditionalists by nature, and it was they who resisted the idea of the new types of car; sleeker in look, softer in suspension, even more comprehensively equipped. Perhaps it was for them that the Wilks Brothers were creating the Land-Rover.

7 *The motor vehicle in maturity* 1946 *to* 1981

215 *Apart from its forward-mounted radiator (made possible by the adoption of independent front-wheel suspension) the 1948 Rover closely resembled its predecessors of 1946–48 which, in their turn, were so like their counterparts of 1940. The sports-saloon seen here was the final development of a shape that was created in the early 'thirties; intelligently modernised in small and subtle ways so that the more conservatively-minded motorists to whom the make had appealed for so long would not be frightened away.*

To some extent the situation faced by the British motor industry in 1946 resembled that prevailing in 1919. There were serious shortages of materials. Factories laid out and equipped for the production of military stores had to be stripped-out and re-tooled before peacetime manufacture could resume. The labour force itself had to re-adjust to a changed set of work conditions, and while the extreme pressures of wartime need were eased, urgency was still of prime importance; demand was enormous. Fuel shortages directly affected power production, and the cuts that occurred interfered seriously with continuity of manufacture. There was, too, the recurring problem of which course to follow; whether to revive 1939–40 models (for which jigs and tools were still available in most cases) and start to make and sell without delay, or wait until the design and development of new vehicles was complete and the necessary production resources had been arranged. Rover was one company that followed the former course [215].

To some extent this dilemma was resolved quite simply by force of circumstances. Under the very different economic conditions of post-war Britain it was essential to make and market as many products as possible for worldwide sale. As far as the home motor industry was concerned this meant that our cars and other road vehicles must lose their former insularity of design; be suitable for use anywhere and under any kind of geographical or climatic conditions. The problems that this raised were tackled without delay, and with an admirable concentration of creative energy. Without waste of time the largest pre-war producers—Austin, Ford, Morris, the Rootes Group, Standard, Singer and Vauxhall—converted their vast plants from armaments-making to the manufacture of certain of their 1939–40 models, little changed except in cost. For the fortunate few in Britain who qualified as 'essential users' and took priority, therefore, a new car was more than twice its pre-war price when Purchase Tax (in effect a wartime luxury levy) was taken into account. It did not apply to cars sold overseas.

In 1945, by Government order, fifty per cent of production had to be exported, supplies of raw materials being continued only if this condition was met. In 1947 the figure was raised to sixty per cent, and then to seventy per cent, and by 1949 it stood at seventy-five per cent. Although there was a worldwide shortage of motor vehicles it was not always possible to take advantage of this fact; Britain, after all, was not the only country with foreign currency shortages and other problems. In spite of all the difficulties, however, the export achievements of the British motor industry in those troubled years were quite magnificent.

Meanwhile, the home market was grossly under-supplied; once essential users had been satisfied there was little or nothing left for the private motorist, who had to make do with his pre-war car until such time as bureaucracy allowed him to buy a new one. For the time being he had to wait patiently and pay occasional tribute to those unknown British workmen who had made his car so well in the first place. Many were still in everyday use in the early 'fifties, often fifteen years old, or more, and they held their value surprisingly well beside second-hand post-war cars, high prices being paid with apparent willingness.

For a few years after the war, the shooting-brake (known also as the station-wagon, utility or estate-car) was made in some numbers, on old and new chassis, many of the smaller surviving coachbuilders finding this unexpected demand a welcome aid in the increasingly difficult conditions they were facing. The type had been fairly popular during the 'thirties in the United States, where Ford was the first maker to offer a production model on the 'A' chassis, in 1930. In Britain, towards the end of the decade, the Ford V-8 station-wagon sold to those who sought a large, fast, inexpensive and rather distinctive load-and-passenger-carrier. Before the war and afterwards, station-wagons were built on elderly chassis, an especial favourite being the Rolls-Royce Twenty. They appeared also on new chassis, not all of which were entirely suitable for the purpose; but these were more easily obtainable than complete cars from the makers concerned (Alvis, Allard, Armstrong-Siddeley and Lea-Francis chief among them) and until the supply position improved what became known as the 'woody' in later years had a run of popularity. It even appeared on a few Rolls-Royce 'Silver Wraith' and Bentley Mark VI chassis; in particular, in a specially-designed model from Harold Radford Ltd. Their 'Countryman' was the costliest and most comprehensively equipped body of its type and time, rivalling in most respects the special cars entered for the Coachwork Competition of the then-important Monte-Carlo Rally.

The obvious practical advantages of the type encouraged one or two of the large-scale manufacturers to offer estate-car alternatives to their standard saloons, the first of these in Britain being the Standard Vanguard. It was followed by the Traveller, on the Morris Minor and Oxford chassis, which has now acquired classic status, and such later models as the Hillman 'Husky', various Fords, and the Mini Clubman. Station-wagon models of the Land-Rover have been available for a very long time and the Range-Rover, at the top end of this market, has set a standard for the world ever since its introduction.

It was not only the supply of new vehicles that was restricted; petrol, too, was rationed until 1950 and high-octane fuel, necessary if the post-war high-compression engine was to give of its best, was not available for another year or two. In the meantime, the British motorist had to put up with Pool petrol because there was no alternative. It is to be presumed that supplies of premium fuel were made available to manufacturers for development testing. This was not the case as far as the leading motor journals were concerned, and it was one of the reasons why so much of their road testing of new cars was carried out abroad, where good petrol and suitable places for performance trials were readily available. For obvious reasons the slower kind of family-car was not to be taken over to the Montlhéry track in France

216 *The $3\frac{1}{2}$-litre Jaguar fixed-head coupé first appeared in 1951, as an XK 120 variant, with 160 bhp; later versions, the XK 140 and the XK 150, differed little in look but had more power. Because racing-car speeds could be reached with such mechanical ease and quietness disc-brakes, which Jaguar had used previously with great success in road racing (with specially-prepared cars) were a very important design element in the production models. The way in which William Lyons shaped this quite small machine was masterly: his control of curves, and the way in which he contrasted their viguour with the straight lines of wheel spokes and window framing, are object lessons for the automobile designer.*

217 *The first totally new post-war Austin was the A40, an up-to-date looking car with i.f.s. an overhead-valve engine, a brisk performance and 'new look' bodywork. The Somerset coupé of 1953 was an interesting alternative to the standard saloon. It was a three-position convertible that had the width, the heating system and the bench-type front seat already taken for granted, even in fairly inexpensive cars.*

for maximum speed evaluation, or to the splendid Jabbeke motor road in Belgium, but such facilities were of the greatest value when high performance cars were being tried.

When *The Motor* tested a Mark VI Bentley in 1947 the car was driven flat-out for nine laps of the Montlhéry circuit (at an average speed of 92.9 mph, incidentally). When the Jaguar XK 120 was tried by the same magazine in 1949 its maximum speed of 126.8 mph was reached on the Jabbeke road. There was nowhere in Britain with such suitable conditions at that time. Brooklands track, which had been used by *The Autocar*, *The Motor* and *Motor Sport* before the war for road test purposes, was no longer available either to those responsible journals or to British motor manufacturers, its wartime conversion to aircraft production being permanent, it seemed. Until the opening of the MIRA test facilities at Lindley, many years later, there was no useable alternative in the country, and cars with a performance as high as that of the XK 140 Jaguar, shown in [216] were taken abroad for proper evaluation.

The fact that so much development testing had to take place outside the United Kingdom was beneficial in itself. It put many manufacturers in first-hand touch with the realities of road conditions and driving habits elsewhere; almost certainly it accelerated the adoption of i.f.s. by those makers who had remained faithful to the beam axle, and awakened others to the fact that future engines would have to be able to run reliably at high maintained cruising speeds. Before the war such matters were of little consequence to the majority of British manufacturers; their products were used in the main on well-surfaced roads and had little chance for sustained high-speed running. Few were taken abroad, and those usually were of the expensive kind. In the 'thirties the Austin Seven, the Ford 8, the Morris Ten or the Vauxhall 14, for example, were unlikely to cross the Channel. Within thirty years this was no longer true: by then the British family-car was being taken abroad on an increasing scale and was well able to cope with road conditions there.

Of the revived pre-war cars only the Standards and the Vauxhalls had i.f.s.; only the Morris Ten and the Vauxhalls were of integral construction—but all had pressed-steel bodywork. By 1948 they had been replaced (Ford excepted) by entirely new models, all of which had i.f.s. The Austins [217], the Morris 'Six', the related

218 Wolseleys, new and very old, in 1948. The all-steel Four-Fifty saloon seen here alongside the first four-wheel Wolseley (of 1899) was a good-looking car, the overall shape of which was an admirable compromise between pre- and post-war ideas on appearance design. Its body pressings were also used for the larger Six-Eighty model of the same make and for the contemporary Morris Oxford and Six.

219 *Immediately after the war the British government of the time tried hard to impose standardisation on the motor industry. The Standard Motor Company was one of the few concerns that took heed, the result being its Vanguard model, which was the only car produced at the Canley works for several years. Although its American-influenced, full-width body attracted much criticism in this country the practical qualities of the car itself made it a considerable commercial success in both home and exports markets. This is the first of a number of full-size 'mock-ups' (developed from much smaller clay models which, in turn, had been built-up from design drawings).*

Wolseleys [218], the Standard Vanguard [219 and 220] and the Vauxhalls had overhead valves; the others continued to use side valves. The engines of the Morris 'Six' and and Vanguard were of the short-stroke type that the ancient RAC formula had discouraged British designers from using for the best part of forty years; since 1946 a flat-rate tax for motor-cars had been in force in Britain instead of the previous horsepower levy, and former restraints no longer existed.

Although it was to be many years before the traditional long-stroke engine gave significant ground to the newcomer there was little doubt that it would do so in the end. The short-stroke unit was far more suitable for universal use; its structure had enormous rigidity and its piston-speed to road-ratio was highly favourable for reliable high-speed running. The number of motorways in Europe was already growing rapidly and it was only a matter of time before construction began in Britain. Until 1909 the engine of the 40/50 Rolls-Royce had been square, its bore and stroke equal, at 112.5 mm. Fifty years later the engine of the Rolls-Royce 'S' series was oversquare, with a bore of 104 mm and a stroke of 91 mm. Apart from their well-deserved reputations for quietness and smoothness of running each was noted for its exceptional durability.

220 *By the time the Vanguard went into production its shape was much improved. It was a 'fast-back', but unlike cars of that type popular for a year or two before the war it was much more spacious inside, with reasonable headroom in the rear and seats now wide enough to take three abreast. It was the first car to be sold by Standard with integral construction, Fisher & Ludlow being responsible for the supply of its body pressings.*

Greatly improved life-expectation became an accepted and welcome feature of the new post-war engines, large and small; due in part to great improvements in design, in part to better use of improved materials and in part to the use of superior lubricants, filtration and lubricating systems. As the short-stroke engine grew in popularity there was a further increase in the interval between major overhauls; the better oils improved matters internally so that cylinder heads need no longer come off every 10,000 miles or so for decarbonising; increasingly, in fact, they stayed put for 50,000 miles or more.

When the Mark VI Bentley and the Rolls-Royce Silver Wraith were introduced in 1946 and 1947, respectively, it was said that the designed interval between major overhauls of the inlet-over-exhaust valve engine common to each was 100,000 miles. That figure was realised in service by many examples. The classic post-war Jaguar engine, the twin overhead-camshaft unit first used in the equally classic XK 120 Super Sports model, and later produced in various capacities and powers, has also proved capable of reaching, or exceeding six-figure mileages before overhaul. These large units, it should be said, were all of the long-stroke kind.

It would be quite unreasonable to expect similar results from a poorly maintained car or from one that was continuously maltreated. The modern engine requires far less attention than its distant forebears but what is necessary should be done at the stated intervals. Until the 'sixties the writer owned, used and maintained pre-war cars built between 1924 and 1937 and found that regular attention kept them going without trouble. Since he became involved with post-war cars he has gradually learned to accept the fact that cylinder heads seldom need to be lifted, that tappets stay correctly set for years on end and that electrical problems are infrequent. Of the eighteen cars owned to date, fifteen have been of British make; all were bought second-hand—more than one with over 100,000 miles of use—and all have been notably reliable, mechanically speaking. Some had costly coachbuilt bodies of composite construction that might have been expected to provide problems after

221 *Years of work normally precede the appearance of any new car and one of the busiest places during the development stages is the factory drawing office. In the mass-production of a product as complex as the modern motor-car, every single one of its thousands of component parts has to be designed, then drawn with the greatest possible accuracy and dimensioned correctly before the manufacturing processes can begin. This scene in the Standard drawing office, with a small-scale model of the then new 8 hp car under consideration by members of staff in the foreground while draughtsmen work on full-size details behind, was taken in the early 'fifties.*

222 *An early Park Ward bodied Bentley. Instead of the usual wood framing, it had a steel frame, the individual parts being welded together in a sequence determined by experiment. The result was a strong, rattle-and rust-free structure, many examples of which are still in service after more than forty years of active life.*

twenty years or more of extensive use. In two cases only (a 1931 semi-Weymann sports-saloon and a 1935 drophead coupé) was there trouble, with wet rot in door pillars and rattling on rough surfaces. The others, a 1930 limousine and 1934 and 1935 sports-saloons, were entirely trouble-free, with no signs of corrosion or metal fatigue in their aluminium panelling or mudguards.

The all-steel body of a 1937 Bentley [222] was entirely rust-free after twenty-five years and, in spite of its type of construction, the doors closed quietly, unlike those of its lineal descendant, the standard steel Mark VI, which clanged shut. The pre-war car provided what might be called cosy accommodation for four, close to the ground and in pleasant surroundings, but with no great margin of head or leg room for rear-seat occupants. It ran very quietly but its ride was firm, in traditional fast car manner, and made speed reductions necessary on particularly rough roads. The post-war car had a far more spacious body, set comparatively high, with ample head- and leg-room. It had the added amenities of radio, and a heating system that was gently effective, and the standard of its interior design was high. So, too, was the standard of finish everywhere, which equalled that of the specialist coachbuilders in every respect. The Mark VI was a very quiet car also, but its ride was much superior in the balance of comfort and stability. There was no longer any question of slowing up on bad surfaces.

223 *The original 1948 side-valve Morris Minor was an excellent piece of mobile sculpture. Its shape overall was simple but comely; its proportions subtle and satisfying; its visual appeal very strong. Although Sir Alec Issigonis is named as its sole creator it is the case that a great many heads and hands were involved in this major post-war achievement of the British motor industry.*

At the other end of the market there were broadly comparable differences between, say, the 1938 Morris Eight and the 1948 Minor. They shared the same willing little side-valve engine and in their different ways had a most attractive personality. The Eight was the straightforward small family-car at its best, by British standards, capable of comfortable 50 mph cruising with a full load on good roads. Although an easy machine to drive, its handling and roadholding were far below the levels set by several small continental cars or by the better kind of native sports-car. The Minor [223], on the other hand, handled with exactness; its steering was light but in no way vague; the car went precisely where pointed and was a joy to drive quickly along winding roads. To begin with it was little faster than its predecessor and took time to reach its 50–55 mph cruising speed with four people aboard. Because of the excellence of its suspension and roadholding, however, it could sustain this rate over good surfaces and bad, without easing up except on the sharpest bends. Another attractive small car of the 'fifties was the Standard 8 [221].

Even higher standards of handling and roadholding were to be achieved in the next major Issigonis design, the equally immortal Mini—like the Minor intended primarily for the family motorist. By 1959, the year the Mini was introduced, the average British car offered very acceptable levels of ride and comfort and roadholding, the art of compromise being well understood by this time. But the cheeky little Mini, with wheels so small that one might be forgiven for thinking that it was running on its brake-drums, handled like a high-grade vintage sports-car, and rode like one, it must be said. In expert hands its ability to travel flat-out regardless of road surfaces and corners made it more than a match for most cars of much higher power and speed, and it was this reserve that made it such a safe car in everyday use, under the management of the average owner.

The Mini [XXIII] was enormously important because of the effective use which its idiosyncratic designer made of space. In the United States there was room and to spare for the large automobile. In the British Isles space was at a premium as far as the urban motorist was concerned. Of a sudden he was provided with a four-seat motor-car that was 10 ft long and 4 ft $7\frac{1}{2}$ in wide, highly manouevrable, tolerably comfortable, highly economical and fast enough for all practical purposes. It seemed too good to be true and there was some initial sales resistance.

The positive virtues of the design soon made themselves known, (helped in no uncertain manner by the imaginative publicity suggested by Laurence Pomeroy, Junior), and if the world did not necessarily become a better place because the flood of Minis began to spread in 1960, motoring for a million and many more was to become a great deal safer, with a degree of enjoyment that the majority had probably not experienced before. In their different ways the Minor and the Mini satisfied the motoring needs of millions, and continue to do so; although production of the former ended in 1970 the number in regular use is remarkably large and low-mileage or restored examples now sell at prices many times more than their modest original cost. Already the Minor is a 'collector's car'—an unpleasant term which places the car with old postage stamps or picture postcards: purchased, put away, peered at from time to time. The Morris Minor lacks many features that are taken for granted on small cars nowadays, but it has one that they far too often lack, and that is individuality. Its handling, its manner of going and its appearance set it apart and give it a special quality. There is no mistaking a Minor, just as there was no mistaking another important British car, the Austin Seven.

No doubt the far-sighted are already salting away Minis as collector's cars of the near future, along with prime examples of the 1100/1300 series cars that were introduced in 1962 with, it was said at the time, a planned production span of ten years. In fact manufacture went on until 1974, and by then almost 2.4 million had been built. ADO 16 (its official Austin Drawing Office number)was the third great design attributed to Alec Issigonis, and it was one of those comparatively rare creations with an immediate appeal. Although it was very much a European car in size and in its use of small-capacity four-cylinder engine with front-wheel drive, its liveliness, handsome appearance, remarkable interior space, and combination of comfort with exceptional roadworthiness not only made it best-selling car in Britain for six years but attracted buyers from all over the world. There was even a special model, the Austin 'America', to meet North American requirements. The Minor and the Mini were very British. The 1100/1300 cars were much more international, reflecting certain continental design practices in the suspension and cooling systems, for example, and relying largely on Pininfarina, of Italy, for their outstanding appearance. In its day there was no better-looking small car built in Britain in quantity and no other that provided so much space for people within such modest dimensions.

224 *The BMC ADO 16 was marketed in a number of forms, the most costly of which was the Vanden Plas 'Princess', first seen publicly at the 1963 Motor Show. On the inside it was a scaled-down luxury car, with fine leather upholstery and door trim, highly-polished walnut woodwork, good quality carpeting and a comprehensive range of equipment. It was very well finished; the sort of small car that many thought Rolls-Royce should be making.*

ADO 16 was marketed under six names: Austin, MG, Morris, Riley, Vanden Plas and Wolseley, and in two forms: the original 1,098 cc model and the 1,275 cc model announced in 1967. Various degrees of engine tune were provided, the Austin and Morris 1100s, for example, beginning life with 48 bhp, single-carburetter units, whereas the contemporary MG 1100 had a twin-carburetter, 55 bhp unit. So, too, did the 1100 Riley, Vanden Plas 'Princess' [224] and Wolseley. Broadly similar changes were rung with the single- and twin-carburetter versions of the 1300, with from 58 to 70 bhp available.

Whatever may be thought of 'badge engineering' of this kind it does make good marketing sense. Brand loyalty can be a decisive selling factor, and the British Motor Corporation, as it then was, had much experience in the matter. So, too, did the rival Rootes Organisation which, since its beginnings as The Rootes Group well before the war, had been more conscious than most British manufacturers of the commercial importance of good appearance design. Those with long memories will remember the stylish little Hillman 'Aero Minx' and the Talbot Ten, of 1936, which were close relatives of the current Hillman 'Minx' (itself a handsome small family-car), and the more costly Humber 'Vogue' sports-saloon. In contemporary publicity credit for its design was given to the couturier, Captain Edward Molyneux; how closely he was involved in fact is a matter for speculation, however. The look of the 'Vogue' dated very quickly. An outstanding feature of Molyneux clothes, on the other hand, has been the timeless quality of their design.

That, of course, matters more nowadays than it did when the investment required to design, develop, manufacture and market a motor-car was a fraction of what it was to become. The nineteen and twenty-year production spans of the Ford 'T' and Rolls-Royce 40/50 were quite exceptional in their time but since 1946 even longer ones have become almost commonplace in Europe. The T-series Rolls-Royce and Bentley models [226 and 227] were phased out in 1980, after fifteen years of

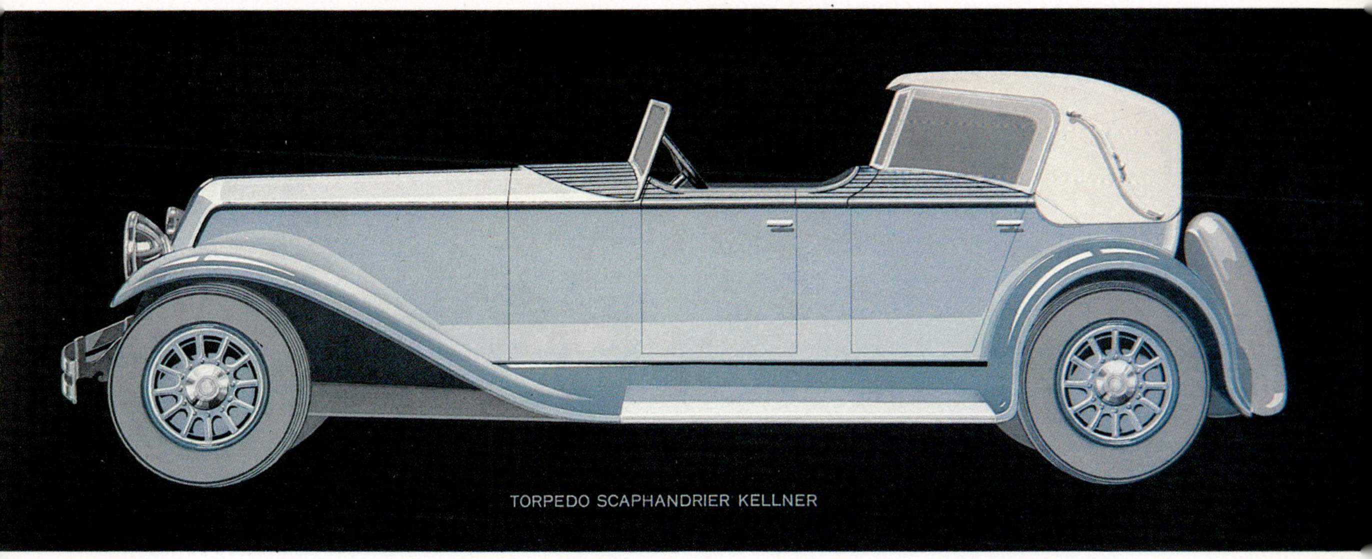

XIX *By 1930 Renault had abandoned their rearward placing of the radiator, putting it where almost everyone else did, at the front. The 'coal-scuttle' prow remained, however, for a little longer, with bumpers as protection. The French Weymann company was responsible for this attractive saloon.*

XX *At the same time Kellner designed this 'Torpedo Schaphandrier'. It has a wooden deck area on the scuttle top and ahead of the rather distant rear seat, and has great style despite the continued use of artillery-type wheels by Renault at such a late date.*

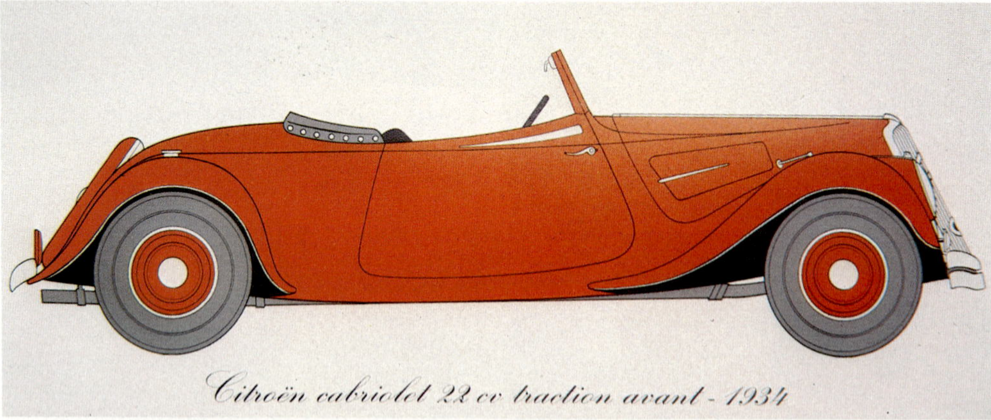

XXI *The 11 hp and Light Fifteen models of the front-wheel drive Citroën are the best-known, and remained in production for thirty years. There was a 22 hp V-8 version in 1934–35, but problems with its engine led to the brief substitution of a Ford V-8 unit, then abandonment of the type altogether. This roadster type of body was not well proportioned but its attractions for the open-air enthusiast were many.*

XXII *W. O. Bentley was responsible overall for the design of the new $4\frac{1}{2}$-litre 6-cylinder and V-12 Lagonda chassis introduced in 1937. They were offered with distinctive coachwork, the standard sports-saloon seen here and the drophead,* plate 209, *proving that curves could be used with taste and to great visual effect at that time.*

226 *Even the manufacturers of the costliest cars have to watch production costs and in the design of the post-war Rolls-Royce and Bentley cars, 'commonisation' of parts, along with great simplicity of design and construction, played an important part. Until* 1955 *these makes were clearly different in appearance but, with the introduction of the 'S' series cars in that year, they shared the same mechanism and the same all-metal body (supplied in shell form by Pressed Steel). Only their radiators, badges and names were different now. This is a* 1967 *'T' series Silver Shadow which went one step further, with integral-construction.*

227 *And this is a* 1977 *'T' series Bentley, little-changed externally (apart from its air-dam, below the hefty bumper designed to satisfy North American legislative requirements) from a shape laid down almost twenty years before. Because of intelligent anticipation on the part of its designers, it did not date in appearance during a fifteen-year-long production life.*

228 *H. J. Aldington himself was the first owner of this 1932 Frazer Nash T.T. Replica—one of the best-looking sports-cars of its day and still highly attractive to the informed eye. Note the graceful curve of the mudguards, the high-set headlamps, the louvred bonnet and exhaust downpipes for quickest dissipation of engine heat, the cutaway sides, the large-diameter speedometer and revolution-counter, the spring-spoked steering wheel and a centrally placed windscreen-wiper motor driving three blades.*

229 *The post-war Frazer Nash T.T. Replica had a much higher performance than its 1932 predecessor and even better handling but it was not nearly so exciting in appearance. In the 1952 form seen here, the visual interest of the earlier car is almost completely absent.*

production and the Mini has passed its twenty-second year. The Volkswagen 'Beetle', designed over forty years ago and in large-scale manufacture from soon after the end of the last war, is still being made in South America.

One costly car that did retain a single identity was the Bristol. It was based largely on the pre-war BMW type 327, one of a small range of German cars sold in this country by the energetic and enterprising Aldington brothers, whose association with the Frazer Nash was of long standing. The latter was very much the traditional English sports-car [228 and 229]—fast, responsive, firmly-sprung, and almost fierce in the manner of its going. It was built by individualists for individualists, and it is unlikely that any two Nashes were alike. The BMW was totally different in most respects; it was made in fair numbers and it was fast, but quiet; softly sprung and comfortable, yet thoroughly roadworthy; and while it, too, was responsive it was at no time fierce. I.f.s and an excellent 6-cylinder 2-litre engine were notable features.

The Bristol Aeroplane Company began in 1946 with a close-coupled saloon closely resembling the 327 body of 1937. Its flowing wing lines and fullsome shapes had not dated; indeed they were followed, though certainly not copied, by Healey (another of the few newcomers to the industry), Jaguar, Austin, Jowett [231], Daimler, Lagonda [232] and others. Before very long they were being used by the specialist coachbuilders; those left in business after the end of the war. The process of

231 *Jowetts, post-war, were socially superior to their willing, homely and generally unattractive forebears, the Javelin saloon and this Jupiter sports model being aimed at middle-class buyers. The long, flowing curves of the Jupiter owed much to those of the BMW and were handled with confidence but little subtlety. Its body was of all-metal composite construction, light alloy panels being attached to a lightweight, tubular steel frame.*

232 *Costlier curves; the car is a 1951 2½-litre Lagonda, which had an excellent 6-cylinder engine, a cruciform-type chassis and independent suspension of all four wheels. Its appearance lacks the distinction of its larger forebears.*

233 *A corner of the body shop at the Bromley works of James Young, in 1961. Special coachwork on several Rolls-Royce Silver Cloud chassis is at the panelling stage of a process of construction that lasted for weeks and called for the services of craftsmen of quite exceptional abilities. Their work demanded manual dexterity comparable with that of the best craftsmen of earlier times.*

shrinkage in their industry had been slow but it had been inexorable, even in the second half of the 'thirties when car production rose year after year, reaching record figures in 1938.

As we have already learned, however, sales of the more expensive cars had not been so satisfactory (perhaps because of the growing possibility of war), and the market for the custom bodybuilder was adversely affected. Barker & Co, one of the oldest and best firms, failed almost unnoticed in 1938, and although its goodwill was taken over by Hooper and its name used afterwards for certain bodies on Daimler and Lanchester chassis, the distinction had gone. It was inevitable; no longer was a Barker body one built almost without regard for cost. Even so, there were enough wealthy customers left to sustain the specialists for another ten years on a reasonably wide scale. Only the coming of the 'sixties meant the going of almost all the once-famous names. At the 1929 London Show fifty-seven different coachbuilders were represented. At Earl's Court, in 1959, there were only thirteen, and of those no more than four were still building on Rolls-Royce chassis. Of the latter Hooper & Co was showing for the last time; H. J. Mulliner and Park Ward had been part of the Rolls-Royce organisation, since 1959 and 1939, respectively; and James Young & Co, which had been taken over by Jack Barclay Ltd many years earlier, was the sole remaining independent [233]. The others had either faded quietly from the scene, were about to do so, or had transferred their interests to goods vehicle or bus bodybuilding, for example, to maintenance and repair work, or to retailing. It was a sad state of affairs, no doubt, but, in the circumstances, inevitable.

There would be no more bodies by Cockshoot or Connaught, Freestone & Webb, Gurney Nutting, Mann Egerton, Arthur Mulliner, Rippon, Thrupp & Maberly, Vincents of Reading, Weymann or Windovers. Not all those makers had been in the top rank but the general standard of their work had been high and individuality had

234 *The Rolls-Royce Corniche is a special bodied version of the Silver Shadow and one of the most expensive production cars in the world. It is a very handsome machine, the size of which is minimised by its excellent proportions, and its standards of construction and finish reach the highest levels. Its drophead body (with power operation of the hood) is made by H. J. Mulliner, Park Ward.*

been widespread. By 1967 James Young was gone and the only major coachbuilder in Britain outside the mass-production field was H. J. Mulliner Park, Ward Ltd fully engaged on the making of very special bodies for the Rolls-Royce Silver Shadow and Phantom V models. Although the latter had a chassis the former did not—its makers had adopted integral construction at last—and the non-standard bodies were modified versions of the standard steel saloon. Nowadays, attention is concentrated on the Bentley Mulsanne, the Corniche [234], and the extremely costly Camargue. The latter is not the aesthetic success that it ought to be, despite, some involvement on the part of Pininfarina.

It is noticeable that the work of designers of his stature suffers badly if it is changed by others, less well equipped, visually speaking. This was the case with the $1\frac{1}{2}$-litre BMC Austin, MG, Morris, Riley and Wolseley cars that Pininfarina was asked to clothe in the late 'fifties. Their basic shape was good enough but overall appearance was spoiled by poor detailing or the intrusion of radiator grills at odds with the rest of the vehicle. Interior design, too, was often ill-matched to the whole.

In the case of the Camargue the shape of the Pininfarina Fiat 132 coupé had been followed; but not closely enough: the car looks to the outsider as if it were suffering from arrested development. It lacks the sheer beauty and distinction of one of the classic designs of the post-war period, the original H. J. Mulliner Bentley Continental [235]. Here was the fast-back of the mid-'thirties properly done; a reminder, if one were needed, that Britain could still turn out a car of high performance and high aesthetic quality. It was extremely expensive. Few were made. In its original form it was held to a strict weight-limit; it was a very fast, quiet and economical motor-car while also being an extremely comfortable one. It was most of these things because it had such a favourable power-to-weight ratio. Unlike many super-cars it was lightly stressed, mechanically speaking, which made its running effortless at high speed and helped to keep fuel consumption low; it also made the intervals between overhauls very

235 *From a number of points of view the Bentley Continental was one of the most significant British cars of the early 'fifties. Less than two-hundred examples of the original design were built, the number of customers worldwide for a four-seat car with limited luggage, very high performance and a price of £4,890 (which meant that the foreign buyer might have to pay close on £7,000 by the time his car reached him, what with transport costs and local import duties) being necessarily small. The Continental is a car that has presence—something that cannot be conveyed by any photograph, however good in itself. This is the H. J. Mulliner design that caused such excitement at the 1951 Motor Show.*

236 *The 911 Porsche has been steadily developed, mechanically and bodily, and at the beginning of the 'eighties stands securely among the finest cars of its time. Its appearance now entitles it to classic status, along with that of an earlier 'fastback', the 1951 Bentley Continental.*

long. There can be few better long-term cars—for use, that is, and not simply for fiscal aggrandisement. In the writer's experience driving an early H. J. Mulliner Continental is one of the really great motoring delights; one that he would be happy to repeat indefinitely on a daily basis. A modern equivalent is the 911 Porsche [236].

By any normal standards, the original Continental was luxuriously trimmed and equipped but many potential owners wanted more interior opulence, not caring, according to Mr George Moseley, whose personal involvement was close, to say the least, whether seats were comfortable as long as they looked it. The level of luxury was raised, and so was the weight; a bigger engine was fitted but there was no increase in performance, and gradually the sleek, fish-like shape gave way to something less distinctive, more orthodox. It was a repetition of what had happened over and over again in that class; weight, and its principal cause, more often than not, the customer, were the maker's worst enemies.

Flowing lines and ample curves were favoured by some manufacturers in the first ten years or so after the war. That these could be kept well under control can be seen [237 and 238]. The bulbous quality of so much pre-war pressed-steel bodywork was retained and was to be found right through the available range of post-war cars, even some of the most costly failing to escape this pernicious influence [239]. Possibly the uncertainties of life at the time had an unsettling effect; too many of the designs lacked clarity of form and decisiveness of line; instead their shapes were clumsy, their proportions displeasing [239a]. Morris and Wolseley cars for much of that decade were of very good appearance, however, their designers acutely aware of the importance of contrasting subtle, shapely curves with straight lines, something they were well able to

237 *The curves of the fibreglass-bodied Jensen 541 were very carefully determined by its designer and because the proportions of the car are so good its appearance has dated little in more than twenty years. It was a fairly large machine and a full four-seater but use of a tubular metal frame and reinforced plastic kept its weight down to 24 cwt, unladen; with a lightly stressed 3.9-litre 6-cylinder engine, of 130 bhp, it could exceed 115 mph and travel at least twenty miles on every gallon of petrol.*

238 *The 'C' type Aston-Martin of 1939 (chapter 5, 187) was quite striking in appearance and was an early example of the breakaway from the well-established form of the English sports-car. By 1951 its successor, the 2½-litre D.B.2 was almost reticent by comparison. The drophead coupé illustrated has an emphatically curved wing shape at either end and the fall of its bonnet top towards the rather pronounced radiator grille is firmly declared. Nevertheless it is a car of discreet appearance overall, its high performance potential scarcely suggested in any way, visually speaking.*

239

239a

240

239 *The 405 Bristol of 1954–55 was an expensive car, of high mechanical quality, and it was available with a standardised saloon body by its makers (right) or a much better-looking drophead coupé made by E. D. Abbott, of Farnham. The bold curves of the saloon were not well managed, the 'speed line' running from the top of the front-wheel arch to the centre of the forward door panel was an afterthought, surely, and the suggestion of a tail-fin plays no useful aesthetic or practical purpose.*

239a *Around 1960 the appearance of most American automobiles was very bad indeed, as this Chevrolet Impala sedan clearly proves. Apart from anything else cars of this kind were so wasteful of road space. Much better things were to come.*

240 *It is obvious that the early post-war Morris, Riley and Wolseley cars were designed by the same team; they share a family likeness and each is very good of its kind. The $2\frac{1}{2}$-litre Riley was especially satisfying and, apart from its decisive, attractive lines, its fabric-covered roof provided an interesting textural contrast to the smooth cellulosed body panels. This was an early use of plastic material on such a large scale.*

do. The pre-war Riley, as made by the Nuffield Organisation, was mediocre- looking. The post-war $1\frac{1}{2}$- and $2\frac{1}{2}$-litre cars had an altogether different look combining modern and traditional features so successfully that they do not seem out of place even now [240].

The difference between traditional and up-to-date thinking was most clearly seen in the Standard Vanguard and the Triumph 2-litre saloons. The former—the car described by Sir Hugh Casson as a 'tin meringue'—was not tall, lean and handsome; it was short and fat, not at all good-looking, with evidence of the 'stylist's' hand

241 *The razor-edge style, pre-war, had been confined to custom-built bodywork. It was cleverly and intelligently used for the Triumph 1800 Town and Country saloon introduced in late 1949—a reasonably priced family car that was made in sizeable numbers. Slender pillars and large areas of glass gave occupants an excellent all-round view.*

everywhere. The Triumph [241] which used the same engine, was aimed at a different market; its inspiration was the 'razor-edge' style used so effectively by Freestone & Webb, H. J. Mulliner, Park Ward and others in the late 'thirties. Its main body shape was an interesting combination of gentle curves and straight lines, with which the bold sweep and decisive shape of its mudguards contrasted effectively.

This style was at its best on the Rolls-Royce Silver Wraith [242], whose tall and dominating radiator encouraged the use of straight lines and flat, or almost flat, surfaces elsewhere, with contrasting curves to give visual interest. It was seen at its best between 1946 and 1950, when a number of variations on the four-light, sports-saloon theme was produced by Freestone & Webb, Hooper, H. J. Mulliner and James Young. Park Ward seemed to be going through a bad spell at that time and produced little of aesthetic merit. Very much later, in 1973, H. R. Owen were responsible for an interesting design [243 and 244], which featured a subtle combination of curves and flat surfaces. Park Ward had been involved in the preparatory work for the new standard steel saloon that was to be mounted on the Mark VI Bentley chassis. It was a direct descendant of the type patented by the company in 1935 and built in such large numbers on subsequent $3\frac{1}{2}$- and $4\frac{1}{4}$-litre chassis. It was a clever mixture of traditional and up-to-date design thinking that was extremely successful in sales terms; with one major change—the fitting of a larger boot in 1952—it was in production from 1946 until 1955, when it was replaced by the bulkier and better equipped 'S'-series cars. From 1949 until 1955 the standard steel body was also available on the current Rolls-Royce chassis (initially as an export-only model), selling as the Silver Dawn [245].

242 *The majority of early post-war coachbuilt bodies lacked the subtlety of form, line and proportions of the best examples from the 'thirties. Standards of workmanship and finish remained unchanged but most of the new cars had a clumsy look. The various elements of this Hooper-bodied Rolls-Royce Silver Wraith touring limousine lack unity and harmony, the strong flowing and falling line of the front wing, for example, having no visual connection with any other part (below).*

243 *Special projects are now rare—but well worth carrying out when the end result is as strikingly attractive as this sedanca, which was shown by H. R. Owen in 1973. It was based on an XJ6 Jaguar base and is notable for the way in which its designer divided the body sides, from front to rear, in the two planes, and for his sensitive detailing of the window areas.*

244 *The interior of the Owen sedanca was a not entirely successful attempt to change and improve the standard Jaguar layout—an almost impossible task because of its already high level of design. It lacks unity and would be improved by a better fit of individual parts.*

245 *The Mark VI standard steel saloon was the first complete car to be offered by Bentley Motors and in spite of its high price several thousand were sold between 1946 and 1953, when it was replaced by the 'R' type, which had a much larger boot but was otherwise little changed. Between 1949 and 1953, the Silver Dawn, a Rolls-Royce version of the Mark VI, was marketed (at first to overseas customers only), mechanically identical apart from carburation, and different otherwise in radiator and hub caps. From 1953 until 1955 the Dawn also had the 'big boot' body. This is a 1954 model, now extremely rare and much sought after.*

243

244

245

246 *A greatly reduced demand for the largest kind of car was satisfied in the late 'fifties and early 'sixties by Daimler and Rolls-Royce and by the few coachbuilders still left. James Young and Co were responsible for this striking touring limousine on a 1963 Phantom V chassis; for such an enormous vehicle their aesthetic success was considerable.*

Standardisation, within reason, had been under serious consideration at Derby from the late 'thirties; a new range of engines had been designed and examples had been undergoing development-testing in France during 1939. Post-war policy was to use the 6-cylinder model for the Mark VI and Rolls-Royce Silver Wraith chassis and the 8-cylinder version for an outsize car, a successor to the famous Phantoms. In 1950 the Phantom IV made an appearance; it was to be almost the most exclusive model of its make ever produced, its sale being limited to 'Rulers and Heads of State'. By 1956, when manufacture ended, only sixteen had been built, and it was not until 1959 that a successor, the Phantom V, was introduced [246]. The latter had a V-8, and not a straight-8 engine, and automatic transmission; otherwise it was basically the same, with a substantial chassis, on to which the largest kind of body could be mounted, and drum-brakes, servo-operated by a system little different from that introduced by its makers in 1924.

Rolls-Royce's involvement in exporting went back even farther, to 1906, in fact, and the company were well aware of the problems of producing cars strong enough to withstand the most trying conditions. Apart from the economic implications a main reason for introducing all-steel bodywork from 1946 onwards was its suitability for universal use, the traditional composite construction causing difficulties in extremes of temperature or humidity and in maintenance and repair far from specialised skills and workshops. Another, of course, was the urgent necessity to produce cars for sale—especially overseas. Pressed Steel produced body parts for the Mark VI (and, later, for the Silver Dawn) at Cowley and put them together there; then the shell was sent up to Crewe, where it was trimmed, painted and finished to the highest standards. For customers who preferred something more personal it was possible to purchase a chassis and command one or other of the coachbuilders to

247 *Before the war only the Light Straight-Eight Daimler had any pretensions to performance, and as its name suggests it was not in the very large car class. The post-war Straight-Eight was, however, as may be gathered from this photograph of a 1948 Hooper limousine; in spite of its bulk and weight it was fast and could be driven rapidly yet safely on winding roads. It was much used for ceremonial occasions and, in old age, for funeral work. This body is interesting because of the sheer size of the rear window and the lightness of the structure above the waist.*

submit designs. Here too, however, there was a certain amount of standardisation (which was not new, even at the costliest levels of the trade), and the number of special, 'one-off' bodies was small. It was very much a matter of falling demand.

The only other large car chassis available immediately after the war were made by Daimler—a 27 hp six and even larger 36 hp Straight-Eight, with wheelbases of 11 ft 6½ in and 12 ft 3 in respectively [247]. The Eight had a reputation for liveliness and roadworthiness possessed by none of its predecessors and while it was in no way a sports-car it seems to have handled like one. It is interesting to recall that in the early years of the century Daimler cars were much involved in competition motoring. Apart from the striking Hooper bodies created at the instigation of Sir Bernard and Lady Docker, which attracted great public and media attention because of their deliberately *outré* shape and decoration, those large Daimlers carried coachwork of uninspired appearance.

The other large-capacity cars of the day were the Austin Princess and Sheerline, with handsome bodywork by Vanden Plas and Mann Egerton & Co respectively, and the Humber Pullman Limousine, a nine-seater with a Thrupp & Maberly body. They served the needs of the Civil Service as far as official cars were concerned and were very popular with car-hire firms and undertakers whose supplies of pre-war Rolls-Royces and Daimlers were beginning to run out. For some time there was a danger that this small but important market would not be properly supplied, once the big Austins and Humbers went out of production, but their place was taken, eventually, by a large, fast and not particularly good-looking Daimler limousine and by a certain number of specially lengthened Fords. For the transport of government ministers and other persons of political, commercial or industrial consequence Humber Super Snipes, the Rover 3-litre and the 3.5-litre V-8 Rover had a long run of popularity. In

248 *The 600 limousine was by far the mightiest Mercedes of the early 'sixties: the super-car of its time; a 300 hp V-8 with petrol injection and automatic transmission that provided luxury transport of a quite distinctive kind. It was extremely fast; it handled like a good small car; it had disc-brakes, power steering and independent suspension all round, with air bags and rubber springs. Much use was made of hydraulics for such internal functions as seat adjustment and centralised door locking.*

an increasingly democratic country they were less obvious than the Rolls-Royces and Daimlers still favoured by royalty and by certain conservatively-minded cities. At a later date the Mercedes-Benz 600 [248] offered the utmost in luxury, comfort, speed and performance.

For the carriage of larger numbers of people the motor-bus continued to serve well, in single- and double-deck forms [249 and 250] and for short- as well as long-distance use. The petrol engine had all but disappeared in this and in the goods vehicle field, except for the smallest types, and the diesel's primacy was assured. Although some engines of the kind had been tried out in car chassis during the 'thirties and had proved to be exceptionally economical their weight, bulk, high cost and noise did not endear them to the average motorist. The only production cars offered with diesel power before 1940 were made by Citroën and by Mercedes-Benz but after 1946 there was increased interest in the practical advantages of the type. In 1950, for example, *The Motor* tested a 1.7-litre Mercedes-Benz diesel that cost only £100 more than its petrol-engined equivalent, and although it was slower and less lively, its fuel consumption of 40 mpg or so—against 25 mpg—gave it a great advantage for the economically minded. For a time there was a diesel option for certain of the BMC $1\frac{1}{2}$-litre cars and its popularity with taxi owners during the past twenty years has been tremendous. For the private owner in Britain, however, it has failed to catch-on in any significant way and such diesel-engined cars as are sold in this country nowadays are of continental origin, Citroën, Mercedes-Benz, Peugeot and Volkswagen sharing the market. Their initial cost is still an obstacle for some; for

249 *The 1957 A.C.V. 'Bridgemaster' was one of the best-looking double-deckers of its time; well-proportioned, unfussy in detail and easy to clean. This type of low-height vehicle is easy to enter or to leave and can be used on routes with unavoidable low bridges.*

250 *Comfort and convenience and a trim appearance are important features of the design of the current Leyland National 2 single-decker. Its driver is given a truly commanding position of control, with a quite unrestricted view forwards, and occupants, either sitting or standing, have an excellent outward view. There is no attempt to disguise the fact that body panels are riveted on to the frame beneath.*

251 *Goods-vehicle design has not altered dramatically during the past thirty years; the forward-control type is still the most widely-used, for all types and sizes. The appearance design of this 1980 Ford is a good example of present-day practice, but it is the extra equipment that is of special interest—the air deflector hinged to the cab roof and the air-dam fitted below the front bumper. These reinforced plastic additions are being used on an ever-increasing scale as inexpensive aids to the reduction of fuel consumption—a matter nowadays of the greatest importance what with rising costs and falling supplies of oil.*

others their inferior performance is a drawback; in practice, however, the saving in fuel costs for the driver who covers high annual mileagues can offset the extra price eventually and the differences in speed and acceleration narrow all the time.

Diesel power is almost universal for goods vehicles, for the same reasons that put it into a preferred position before the war. Gardner and Perkins are still among the leaders in an industry that has given Britain world importance since the early 'thirties, along with a very large overseas trade. In general there has been no dramatic outward change in the look of the average lorry or van since the war; the forward-control (or cab-over-engine) type still predominates [251], on rigid and

XXIII *The basic shape of the Mini remains unaltered but over the years discreetly different versions of the standard car have been successfully marketed, this 1980 special-equipment model being a good example. It has an opening roof, special seat-trim, special wheels, a distinctive external treatment, scarcity value because of its strictly limited 'edition', and a price a good deal above that of the normal model.*

XXIV *The Ford Fiesta Ghia is a good example of the present-day small saloon car (in this case a 'hatch-back') that performs with the vivacity of a sports-car yet returns good fuel mileage; careful attention to its shape has increased its aerodynamic efficiency to such an extent that no excess of engine tuning is necessary. The Ghia version carries a comprehensive range of equipment and is top model of the present Fiesta range.*

XXV *At one time, manufacturers tended to keep their experimental projects hidden away, but in the post-war period there has been a refreshing willingness to show off design ideas—often at one or other of the more important international exhibitions. In 1978 Vauxhall Motors displayed their Equus, a sleek two-seat sports-car powered by the current 2.3-litre engine that stood a better chance of eventual production than many projects of its type. Changed economic circumstances, however, have almost certainly stopped this.*

XXVI *Fifteen years after the introduction of the Silver Shadow and T-series Bentley, Rolls-Royce introduced new models. Here is the Rolls-Royce Silver Spirit saloon, an up-to-the-minute vehicle of total luxury that is built to traditional standards of excellence by making best possible use of the capabilities of craftsmen and machines.*

252 *The modern motor lorry travels greater distances in shorter journey times than was ever possible before the introduction of motorways, and in very much busier, more arduous conditions. The importance of making the driver's task as easy and untiring as possible is a prime design consideration, therefore, and cab amenities as comprehensive as these are taken for granted now. This is a 1980 ERF interior—a sleeper model with curtains, reading lights, an internal power point, a cigar lighter and larger map storage area. Quietness, comfort and safety are special features. The cab itself is made of fibreglass.*

253 *The driver of the 1980 Leyland Roadtrain sits in great comfort and in a commanding position, provided with instruments and controls arranged for maximum ease of use and well-insulated from the big diesel engine that shares the cab with him. It is a far, far cry from the unheated, uninsulated, windscreen- and instrument-less cabs of earlier times.*

articulated vehicles alike. Where there has been a significant change is in the provision of comfort and convenience for driver and crew [252 and 253]. Relaxations on size, tare weight and load-carrying capacities have enabled manufacturers to produce some extremely large machines that can maintain cruising speeds above 60 mph on suitable roads, and of these there are more and more, even in this country. Since the opening of the first stretch of the M1 motorway in 1959 and subsequent extensions to Britain's network of modern, high-speed roads, there has been an enormous reduction in journey times for all types of vehicles.

254 *Before motorways it was seldom possible to make full use of the performance of the motor-car for any length of time, especially in the British Isles. In Germany and Italy, however, where roads of this kind had been built before the war, much had been learned about designing and making vehicles that could stand up to prolonged high-speed operation reliably and economically. The Ferrari is one of the most celebrated Italian makes of the post-war years, with high top speed and accelerative powers, splendid handling and memorable appearance. Carrozzeria Pininfarina of Turin, which has been the most prolific source of new ideas in car-body design for more than fifty years, created this singularly beautiful occasional four-seater—or G.T. coupé—on a 330 Ferrari base (above).*

255 *The Lotus Elan was an English car that had the mechanical stamina to cruise at high speeds on motorways, when necessary, and the liveliness of acceleration and very high standards of roadholding and general control to make fast cross-country journeys on inferior roads as much of a pleasure as ever they can be in present-day traffic densities. The first Lotus to have special closed bodywork was the 1957 1,100 model and it was designed by the Italian coachbuilding firm of Ghia. The Elan was all-British; a sleek, neat little car with a fibreglass body, retractable headlamps and a tiny radiator air intake set below the front bumper (above, left).*

256 *The designers of the DS19 Citroën followed their own ideas to a logical and totally distinctive conclusion. In production form the car was at least a decade ahead of its time, mechanically and aesthetically speaking. Powers of anticipation are essential in an industry so closely linked to fashion, and an ability to look so far forward was yet another of many Citroën achievements. Note that the radiator has disappeared almost completely as a visual feature (above, right).*

Passing mention has already been made of the effect that changed driving circumstances had on vehicle design [254 and 255]; engines for the motorway age had to be capable of holding high speeds for hour upon hour without mechanical failure, for example, and without excessive consumption of fuel. This was something that Italian and German designers had learned about before the war, after the

259 *The shape of the Land-Rover is functional and workmanlike; not at all self-conscious. An important visual feature is the pronounced tread-pattern of its cross-country tyres.*

260 *The Range-Rover can cope with cross-country, off-road motoring with ease and with greater comfort than that provided by its sister Land-Rover. Unlike the latter, it was designed for main road and urban use as a load and passenger carrier and great attention was paid to its appearance design. Here again boldly-patterned tyre treads have visual importance, as do the distinctive wheels.*

opening of their new, fast motor roads. Braking and suspension also had to be improved to meet new and more demanding conditions and this accelerated tyre development as well. The gradual introduction of the radial tyre was very much a European matter; the type is now predominant on this side of the Atlantic, and with good reason, the extra cost being offset by a life two or three times longer than that of the conventional crossply tyre and by greatly improved performance. Nowadays, tyre design can be an extremely important factor in the design of the car as a whole, a fact well exemplified by the advanced Citroën DS [256], of 1955, which relied heavily on the Michelin 'X' tyres with which it was equipped, and the Jaguar XJ6, of 1969, for which Dunlop designed very large low-profile radial tyres. No one who follows motor racing can fail to have noticed the great importance of using the right kind of tyre to suit prevailing track conditions; one kind for wet roads and another for dry, with subtle variations of each. For road use, the tyre designer has to make many more compromises of a practical kind and he has to take cost into account, as well as the appearance of the tyre, which can be an important visual feature in certain designs [259 and 260].

In a book of this length, covering almost one hundred years of motor vehicle development, it is quite impossible to deal with all aspects of a vast subject, let alone each significant model ever made. This is particularly true of the post-Second World War period, during which the number of individual manufacturers has been reduced, what with failures or mergers, but the number of vehicles made has increased quite enormously. In looking back over thirty years or so, and specifically at the state of affairs in Britain, one is very much aware of the increased importance of the foreign car in the overall picture. After the 1918 Armistice, American, Belgian, French and Italian vehicles were soon on sale in Britain, but it was not until the end of the 'twenties that Mercedes-Benz, for example, made an official appearance at the London Motor Show, although cars of that make were certainly in the country long before then.

After the Second World War dollar shortages kept out American cars for a long time; those seen were owned by diplomats or visitors [261] in Britain. Citroën and Renault had established assembly plants well before the war and were able to resume production without much delay. Italian [262] and German cars were available by the early 'fifties, the Volkswagen, of course, attracting a cult following soon after its practical merits had been proved in everyday use [263]. Dr Ferdinand Porsche, its designer, chose to use air-cooling because of the practical advantages in a low-cost, go-anywhere car: among them simplicity of manufacture and assembly, and freedom from freezing-up problems in extremes of climate. As it happened, the wisdom of his decision was confirmed when the Volkswagen became the second universal car, the successor to the Model 'T' Ford, in the 'fifties and 'sixties.

Like the Ford it was very well-made, of first-class materials, and parts. Maintenance and repair were cheap and available almost anywhere in the world. Although the Volkswagen body was not noted for space its capacity was sufficient and in an age of improving roads the fact that it could be cruised indefinitely at its maximum speed was not simply of academic interest. Air-cooling had been popular for a time after the Armistice; almost thirty years later it was back in fashion on a large scale.

261 *The appearance of most American cars of the 'fifties and 'sixties was poor by any standards, regardless of first cost. It is hard to believe that the Cadillac was top American car from the visual evidence here; since 1958, however, there has been great improvement. Space exploration was still some way off when this car was designed but rocket shapes featured: the front bumper over-riders look like nose cones and the inspiration for the tail-fins came, surely, from the rocket.*

262 *Eventually American design was favourably influenced by the Italians whose motor-cars were of uncomplicated shape; dependent on excellence of proportion and a correct relationship of parts for their effect. This was as much a feature of standard production models as it was of special cars like this superb Pininfarina Coupé Speciale on a Fiat 2300's base. It is a very beautiful machine.*

263 *Body and mechanism are joined on the production line at the gigantic Volkswagen plant at Wolsfburg, in West Germany. Assembly of such a complex piece of machinery as the modern motor-car is, in itself, an enormously complicated matter, dependent on meticulous planning beforehand and high, maintained standards at every single stage of production.*

It was a feature of another unconventional but essentially practical car, the 2cv Citroën [264], brilliantly conceived as peasants' transport, of the simplest and cheapest kind; practically indestructible, superbly comfortable, incredibly inexpensive to run and quite beyond fashion. It was, one feels, deliberately ugly. It was even more of a go-anywhere vehicle than the Volkswagen 'Beetle' and it was the sort of basic transport that only France, and only Citroën, could produce; so idiosyncratic, so truly original. It had no English equivalent. Its rival, the Renault 4 [265] is of more generally acceptable appearance.

Until well into the 'fifties in Britain the simplest mass-produced cars were the tall, narrow and distinctly audible Ford Anglias and Prefects, of largely pre-war design, that bucketed their way along (in anything but a straight line as a rule). They were reliable, economical and cheap to buy; in 1951, for example, the 8 hp Anglia cost a few pence over £479 and the 10 hp Prefect a little over £572, with tax. Durability was an outstanding feature.

Long life of another kind distinguished the post-war front-wheel drive Citroëns, assembly of which had been resumed at Slough without delay. They had been introduced in 1934, or, rather, their 12 hp forebear had, and fifteen years later their lead, in some respects, was still considerable. From the start, integral construction of body and frame had been an important feature, as had torsion-bar suspension all round, and the low set of the car, with a wheel, quite literally, at each corner, gave it a look of stability confirmed on the road. By contemporary standards its comfort of ride and cornering powers were unmatched, and they are still impressive despite the very great improvements made since 1934, notably by Citroën themselves.

In 1969 the Jaguar XJ6 and Daimler Sovereign [266 and 267] set new standards in these respects, however, and in quietness of running; it is doubtful if there is anything superior, even at the time of writing. Isolating mechanical noise from the

264 *Like earlier utility cars, the 2cv Citroën was meant for work and not for show; good appearance was of no consequence whatsoever, and to say that it had to be peasant-proof is no more than a statement of fact. Its tiny 2-cylinder engine made as much noise as a larger one and could take car and occupants along, in superb comfort as far as ride was concerned, at 40 mph. Body features included a roll-back canvas roof, easily detachable doors and extremely comfortable hammock-type seats.*

265 *On the outside, the Renault 4 that went into production in 1961 was of less controversial appearance than its rival, the 2cv Citroën; box-like, homely, essentially practical. Its interior—like the rest of the car—has remained almost unchanged for almost twenty years. It offers adequate room for four and luggage, seat comfort is remarkable, an opening roof is an optional extra and a tailgate is standard equipment.*

266 *The excellent suspension of the XJ6 Jaguar and Daimler Sovereign is complemented by comfortable seating: their fine appearance on the outside is matched on the inside. This is the front compartment of a 1970 Sovereign, with central console carrying the automatic gearbox lever, ashtray, cigar lighter, air-conditioning controls and radio, and a comprehensive range of instruments and switches.*

267 *After more than ten years in production, the XJ6 Jaguar shape is largely unchanged and there seems no real reason why any alteration should be made for some time to come. Its basic shape is a simple one but fine proportions and precise control of detail raise this classic car far above the average. The car illustrated is a 1970 model.*

264

265

266

267

occupants of cars is no longer a mystic art but suppressing wind- and road-induced sound, (especially the latter), is another matter altogether. High standards were possible in cars with separate chassis, a good example being the Rover in its P4 form. With integral construction it was a lot more difficult, and with the introduction of radial tyres almost impossible, it seemed, at first. But the Peugeot 404 showed the world motor industry that road noise could be almost completely excluded, and many examples were sold to rival manufacturers who hoped to discover the secret by experiment and dissection.

With such standards of comfort and roadworthiness now prevailing it would be unfair to single out more cars for specific mention; the Jaguar [267 and 268] has been mentioned because of its supremacy among British makes and the Citroën because of its supremacy overall, road conditions in France compelling manufacturers over the years to develop suspension systems of uncanny efficiency. The matter is familiar to the Italians also; Fiat, like Citroën, are especially good at making small cars [269] ride with the smoothness of much larger ones—a lesson not yet learned by the makers of the Mini, one of the best-handling small cars in the world but by no means the most comfortable. Some present-day sports-cars have a much better ride.

Sports-cars have changed greatly since 1946. In that year the MG Midget was back [270], looking much the same as it had done in 1939, when the war ended production. The 'TC' was a minor motoring masterpiece as far as its appearance was concerned and although its successors, the 'TD' and 'TF' models, lost some of that quality their sales success, particularly in North America, was proof that Abingdon

268 *The special qualities of quiet running were not confined to the Jaguar XJ6 saloon, which was designed as a top-quality luxury car of high performance. Its sister 'E' type was an out-and-out sports-car, with exciting performance and exciting appearance as prime features. Unlike the traditional sports-car however, this modern speedster combined racing-car speed, acceleration and roadworthiness with saloon-car quietness, comfort and amenities. These are 1971 models—a coupé and roadster, respectively—built to special USA specification.*

269 *The first really handsome small car was the Fiat 500, which sold in millions on an unassailable reputation for reliability and economy of operation. In 1969 this de luxe model was launched, at a total cost in Great Britain of £515/7/6d, including Purchase Tax. Apart from its opening roof, which was standard equipment, it had such extras as reclining seats, carpets, properly trimmed door panels and a fuel gauge. The standard model, it will be gathered, was starkly equipped.*

270 *The classic English sports-car in its final form—an MG Midget of 1949. It is a timeless design, of great quality, its appearance developed and refined over many years of production.*

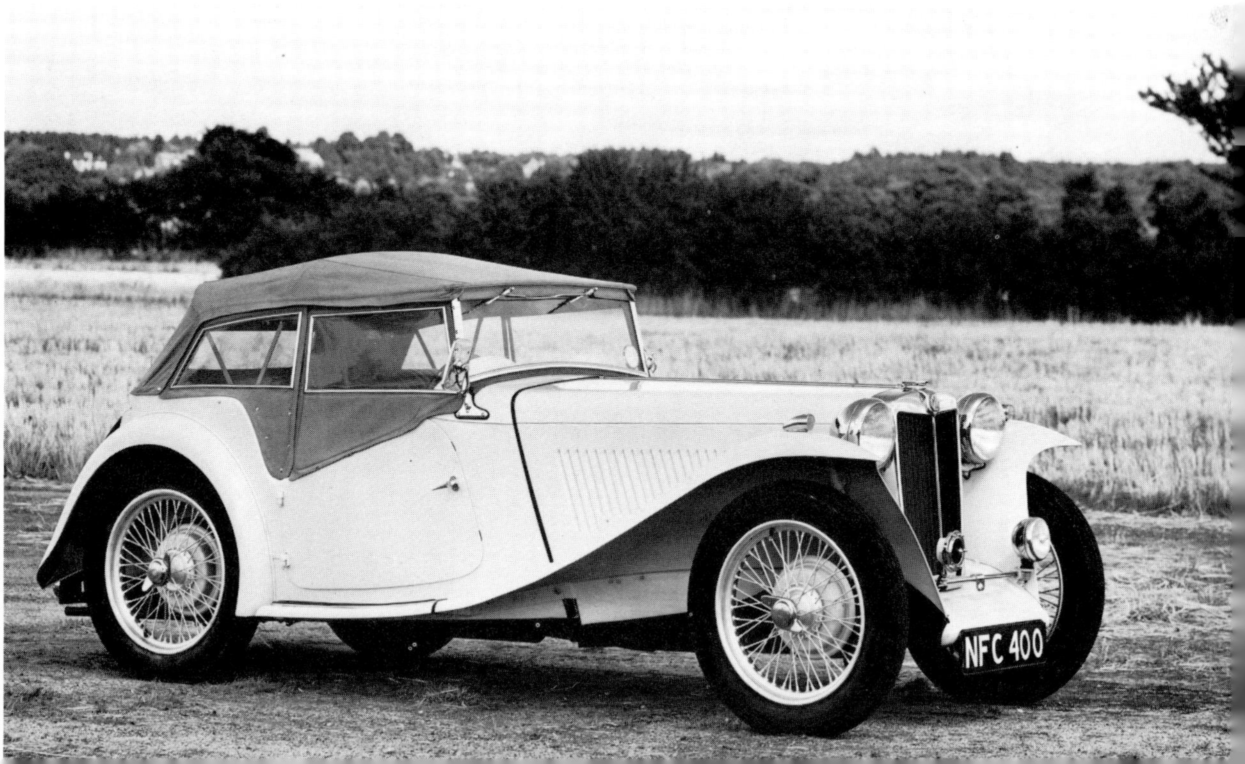

had judged the mix of old and new features very accurately indeed. Even when the pre-war shape was abandoned altogether and the all-enveloping 'MGA' introduced, demand was not affected. The post-war sports-car owner, it appeared, preferred such modern conveniences as i.f.s., less obtrusive mechanism, interior heating, radio, comfortable suspension and good weather protection [271 and 272].

MG did not have the market all to itself, however. Singer, its strongest rival before the war, was fading from the motoring scene, and had no proper sports-car in production. There was powerful competition from the little Austin-Healey 'Sprite' which, in its original 'frog-eye' form, first appeared in 1958. Its BMC 'A' series engine was a two-carburetter version of the unit then used in the Austin 'A35' and Morris Minor models, and it was used to propel a shapely car weighing little more than 13 cwt, at 80 mph. At £668 it was excellent value; a year or two later an almost identical MG Midget version appeared and it, too, was a great commercial success.

271 *This 1977 MG photograph shows the essential visual differences between the old- and new-style sports-car. Each in its day—the 1952 'TF' and the 1977 MGB GT—was a thoroughly satisfactory solution to the design problems involved, a point often forgotten by those who try to make direct comparisons of the unlike.*

272 *A great many sports-cars have close links with production saloons, and in motoring history this is no new thing. Nowadays, differences in performance and general handling characteristics are less marked but visual alterations can result in exciting looking cars. This set of photographs shows an advanced sports-car, the Ghia Corrida, that was designed by Ford Ghia Operations and displayed at the* 1976 *Turin Show. It was based on the floor pan and running gear of the Ford Fiesta* 1117 *S and its coupé body had gullwing doors and a two-piece tailgate. It is an example of the kind of project that is shown publicly nowadays even though it may never reach production.*

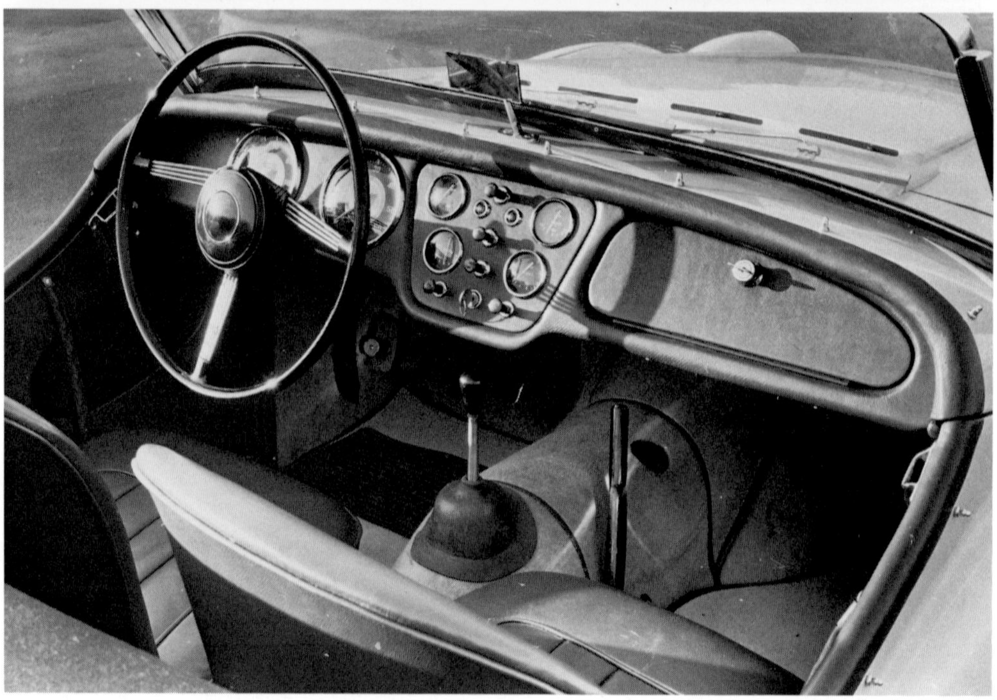

273 *The Triumph TR2 was a small, swift, simple and relatively cheap sports-car that was designed in a hurry around existing mechanical components, as much as possible. How fast it could be made to travel was shown at Jabbeke, in 1953, when this specially-prepared car exceeded 125 mph. The cockpit cover and seat-surround were made of fibreglass—an early example of the use of this lightweight material by the motor industry.*

274 *As more and more British cars found foreign buyers it was essential that conversion to left-hand drive could be easily and economically effected in production. The TR2 cockpit is typical of the way in which this problem was tackled; steering wheel and column, and speedometer and rev. counter could be installed on either side of the car.*

275 *Straight lines and gentle curves contrast well in the 1980 Triumph TR7 drophead. Safety feature include the massive rear bumper, the prominent rear lighting, head-rests, a cockpit as free of potentially dangerous items as possible and a windscreen frame that doubles as roll-bar.*

Then there was the Triumph TR2, of 1953, with its 100 mph-plus maximum speed, high fun-factor, (an important element in this kind of car), and low petrol consumption and first cost [273]. In 1954 it was the least costly car of its speed available in this country and in many others also, at around £900 with tax. Not surprisingly the TR2 [274] sold well, as did its successors. By the time the TR6 appeared, in 1960, the car was a good deal more powerful and complicated than the straightforward, Standard Vanguard-engined original, and it was, of course, a great deal more expensive. It had a 2.5-litre fuel-injection 6-cylinder engine, a rather bland appearance and excellent amenities. Production ended in 1976, by which time its successor, the TR7, had been on the market for a year. The latter, in fixed-head coupé form and, since 1979, in a convertible version [275], has been a popular car on the North American market where its bold styling compels attention. Whether or not the even more forward-looking Vauxhall Equus sports-car shown in 1978 [xxv] will reach production is doubtful, however.

275a *During its extended production life the appearance of the DS range of Citroëns changed little, which was part of its maker's policy. Although the basic shape of the Rover 2000 and 3500 cars also remained unaltered in more than ten years of manufacture what is known as 'facelifting' was allowed for at an early stage of the design process. Here is the 3500 in its original, 1967 state.*

275b *For 1971 the 3500 appeared with a neater front-end. David Bache, the designer, made effective use of plastic for the grille and for the paired headlamp surrounds.*

Another English sports-car of totally different type, the Morgan V-8, also continues to sell well, here and overseas, despite (perhaps because of) its archaic look and old-fashioned sports-car ride. Its front suspension design dates back to 1910 or thereabouts, when F. S. Morgan, the founder of the firm, first used his sliding-pillar type of independent front-wheel springing. The engine used nowadays is the splendidly smooth and potent Rover V-8, a discarded Buick design that has been most successfully revived at Solihull, and the combination of unobtrusive power with light weight and highly individual handling makes for a very special kind of motoring. The same engine gave splendid service in the now-discontinued Rover 3500 [275a, 275b] and is still used in the current SD-series car [275c] and a Land-Rover variant.

The fact that many modern saloons are faster than sports-cars proper is of no real consequence; after all, present-day lorries and buses can travel at least as quickly as many cars. In each case, increases in aerodynamic efficiency resulting from more

275c *It is highly unlikely that there are 40,000 chauffeurs still employed in Britain, but suitable cars are still available for them to drive. This 1980 Rover V-8 has a division between front and rear seats, an opening roof and a hatchback.*

276 *Appearance can go for nothing when performance is of maximum concern. Somewhere beyond the air-dam, the air-intakes ahead of the rear wheels and the tail wing there is a Ford Turbo Capri. Front and back wheels are of different diameters on this rear wheel drive flier which, in the manner of the day, is a nimble, mobile advertising hoarding.*

277 *The solid, but far from stolid Saab 99 first appeared with a Coventry-built Triumph engine, and although the currently-used 2-litre unit is now produced in Sweden, its design is largely based on that of the English original. A major feature of the 99 is the attention given to the security, safety and comfort of users.*

278, 278a *The Volvo P1800 sports-coupé, as another example of international collaboration. Although most of its mechanism was produced in Sweden, its carburetters and its overdrive unit were supplied from England, by SU and Laycock respectively, and its sleek body shell was made by the Pressed Steel Company at its Linwood plant, in Scotland.*

scientific use of air-dams [276], 'wings', and so on, and such boosts to engine output as turbo-charging play a vital part and are now becoming a commonplace. However fast the open car may be it will always appear to be travelling even faster because of the exposed situation of its occupants, and that is a great part of its appeal for many owners. For others, the almost complete isolation from the outside world provided by the average closed car nowadays, is its attraction. Whether one wants to be part of one's surroundings or not is an entirely personal matter; fortunately it is still possible to choose, and to choose, moreover, from a range of vehicles better overall than at any previous period in motoring history.

Since its beginnings the design and manufacture of road vehicles has been an international, not simply a national, matter [277 and 278]. Ideas and the originators of ideas have crossed frontiers freely; the cross-fertilisation process has always been at work; the traffic not all one-way. In spite of better communications, in spite of tighter-knit communities like the Common Market, for example, and in spite of the bureaucrats who would like to see us all crawling about at a low fixed speed inside invulnerable structures, national characteristics prevail to a remarkable degree, the

277

278

278a

279 *The Vauxhall Royale of 1980 is of largely German origin; in effect it is an Opel, made by a General Motors company, and marketed in Britain by another General Motors company. It is also a fine-looking car, with the quietness, the speed and the very complete range of amenities expected in this, the executive class. Power steering, electrically-operated windows, a sunshine roof, headlamp wash/wipe and stereo-cassette/radio are supplied as standard fittings. The sloping prow is a current design fashion.*

motor-car remaining as a mobile reminder of the individuality of man. And this despite the fact that not all the components of a car may have been manufactured in the same country, let alone the same factory [279, 280 and XXIV].

Since 1946 one has seen the average British car change in its suitability for international use. American influences on car appearance have given way gradually to the less exuberant but aesthetically superior ideas of the leading Italian designers on a worldwide scale. One has seen the continuing attraction of the British style sports-car, on road and track; the ups and downs of national fortunes in motor racing at home and abroad; the reduction in the number of makes and the increase in the number of vehicles made; the slow spread of motorways and the rapidity with

280 *The Opel Monza coupé has a 3-litre, 6-cylinder, fuel-injection engine that produces upwards of 180 bhp. Its body shape was carefully developed in wind tunnel tests, two positive results being a top speed of over 130 mph and a government-tested fuel consumption of 25 mpg at a constant 75 mph. It is a hatchback of exceptional beauty of appearance.*

which cross-country journeys may now be made. One has also seen how much attention has been given to raising levels of comfort since 1946. Before then, for example, few cars made in this country had any kind of heating or properly-designed ventilation system, and the urgent necessity to produce cars suitable for the extremes of world weather hastened development. Smiths (until then principally famous for their car instruments) conducted extensive development testing of their new heating systems in Scandinavian winter conditions.

The stage has now been reached at which the average car, and not simply the special, more costly de luxe model, is richly equipped: a trend accelerated by the attention given to this matter by Japanese car manufacturers [283a]. Nowadays the

281 *The shape of the current British Leyland Princess 1700 HLS can be traced back to some extent to the design of the original Saab 99, in the late 'sixties, which was almost the first saloon to feature the falling-forward bonnet line and the raised rear. The English car is, perhaps, the best-looking medium-price model presently available from its makers.*

282 *Plastics predominate in the interior of the Princess 1700 and 2000 models—door facings, fascia mouldings, console, seat coverings and so on are shaped from this versatile material. For such a reasonably-priced car, the equipment is remarkably comprehensive. Among features taken more and more for granted are a stereo radio (with speakers located in the door trims), heater outlet grilles and eye-level ventilators, a large, lockable glove compartment, a console between the seats (with safety-belt clips at its rear) and a two-spoke steering wheel, the design of which takes driver safety into account.*

options tend to be real luxuries—costly stereo-cassette and radio installations, for example, complex and highly-efficient air-conditioning systems, with refrigeration for extremes of heat, or special external paint treatment. The 1979 British Leyland Princess, seen here in 1700 HLS form [281 and 282] is an example of the standard of design set in a moderately priced car. Its shape is a good and positive statement, basically simple but visually interesting. Note the use of decorative lining above the shaped wheel arches, the distinctive wheels and the vinyl-covered upperworks.

283 *Nowadays the estate-car exists as a distinctive vehicle type, giving away little to the saloon in performance, appearance or interior amenities. The 1980 Ford Granada Ghia is a good example of the high standards achieved. It has a 2.8-litre V-6 engine, which spirits it along quietly and quickly, almost regardless of load. When it is used as a two-seater, 77 cubic feet of luggage space is provided. With five occupants aboard that figure is reduced to 41.7 cubic feet.*

The estate-car has come into its own [283]; no longer a potential source of boom and rattles but a vehicle as quiet and well equipped as its saloon counterparts. The Morris Minor Traveller, which appeard in the early 'fifties with external wood frame members as a low-cost utility, has now achieved classic status and original examples, or good restorations, fetch surprisingly high prices. It is certainly one of the best of all post-war cars to own and drive. There is every chance that its stablemate, the Metro [284] will merit the same reputation.

283a *The Japanese Honda Accord is notable for the extremely comprehensive nature of its standard equipment and for good road behaviour. Externally, however, it reveals no positive national characteristics.*

284 *The Metro is the British Leyland challenge to the other small-car makers of Europe, a distinctive-looking little hatchback of extremely ingenious design.*

The writer has also been an interested and involved party in the use and preservation of the older car. This is not at all a post-war 'exclusive'; the Veteran Car Club (VCC) of Great Britain was founded as long ago as 1930, and the Vintage Sports Car Club (VSCC) is its junior by only a few years. Before the First World War the *Motor* magazine had gathered together a number of very early vehicles for display, a most laudable project which came to nothing because of the outbreak of war in 1914. Public interest of a localised nature had been sustained since 1927, when the first London to Brighton Commemoration Run was held, and, in any case, the number of elderly vehicles still in use was surprisingly large.

This state of affairs persisted after the Second World War. Many cars of the 'twenties and 'thirties were being driven into the 'fifties because their owners were still able to make daily use of them. During the Second World War, as we have seen, there was a build-up of interest in cars of earlier times and this interest actually increased with the return of peace. Some keen motorists did not care for the way motor-car design was going (i.f.s., for example, was suspect to some) and turned to the past, to the 'twenties

285 *With rising traffic congestion in urban areas and falling supplies of fuel, manufacturers are actively investigating smaller and much more economical kinds of car. This is an experimental design from Ghia Operations, of Turin; a suggestion for a compact five-seat city car, based on the Ford Fiesta, with engine and transmission moved to the rear. Note the ingenious, space-saving luggage space in the doors.*

in the main, for personal transport. Others, more open-minded, perhaps, but impecunious, could afford a fifteen- or twenty-year old Austin Seven, MG Midget, 12/50 Alvis or 14/45 Talbot and make do with it quite happily, until funds and improved supply allowed them to indulge in something up-to-date [285].

The words 'veteran' and 'vintage' were much-used. Cars of the 'thirties were looked down upon in general, though exceptions were made by the VSCC in the fullness of time—the Rolls-Royce-built Bentleys were officially recognised as 'post-vintage thoroughbreds' whereas the Anglo-American Railtons were not, to begin with. Veterans were cars made before the 1 January, 1905, and as such were eligible to take part in the annual London to Brighton Run, but the VCC recognised all cars made until 1916 (latterly 1918), regardless of type or quality. Those pioneering organisations are internationally recognised and respected.

Until the beginning of the 'sixties the sight of vintage cars was taken for granted, more or less. Since then, however, natural wastage has reduced the number of survivors; rarity value (not necessarily in monetary terms) has increased, and few cars of that kind are used for daily transport nowadays. Interest is at a higher level, perhaps, than ever before and a new, minor industry has grown over the past ten years or so in finding, restoring and marketing the older car of special interest or quality [286]. The days of the great coachbuilders may be over [XXVI], but in recent times the number of small concerns concentrating on body repair and reconstruction or building from scratch, on veteran and vintage chassis and others of later date, has grown to a number close on one hundred in the British Isles. Another hundred specialise in hood-making, painting, replating, upholstering and trimming, lamp-making and restoring, repairing and servicing instruments of all kinds, the supply of accessories of correct type and of tyres, brake linings and other expendables, retailing,

286 *It is still possible to find craftsmen whose skills match those of their great forebears. Here Bob Gill, of Marsh Developments of Hickstead, is seen with the partly-finished ash-framed body of a 1921 Angus-Sanderson that he was making for Mr Michael Worthington-Williams. It is a faithful reconstruction, copied from the only other car of this make still running in Britain.*

transporting and so on. The one make clubs thrive, and while their appearance may not always meet with hearty approval from purists, 'custom cars' have become more and more popular with a certain section of the enthusiasts' movement and prove that traditional standards of workmanship and finish can still be maintained. The growth of the motor museum as a nation-wide attraction in post-war Britain may be attributed to the single-minded enterprise of Lord Montagu of Beaulieu, whose National Motor Museum is the model of its kind, the *nonpareil*.

On road and circuit, in our motor museums, on film and in the pages of magazines and books we can see what our great motor industry has achieved since its tentative beginnings in the late nineteenth century. For all that time, and longer, the Institute of British Carriage and Automobile Manufacturers (and its predecessor) has been at the heart of this activity. At its centenary, the prospect of the next hundred years, in an oil-starved world, is an intriguing one to say the least.

287 *The restoration of veteran and vintage cars is a matter to be taken very seriously indeed but those responsible for the 'custom' or 'fun' cars that have grown greatly in popularity in this country in recent years, long after their development in the United States, are able to take greater liberties. The designer and owner of this elderly Mini van has given it a mottled paint finish, strongly-shaped bonnet and mudguards, an air-dam, old-fashioned running boards, a 'wing' on the roof at the rear, and a structure on its bonnet-top that seriously reduces forward visibility.*

Select Bibliography

William Boddy *The History of Brooklands Racing Course* (1957)
W. Worby Beaumont *Motor Vehicles and Motors* (1900)
A. Bird and I. Hallows *The Rolls-Royce Motor Car* (1964)
A. Bird and F. Hutton-Stott *Lanchester Motor Cars* (1965)
A. Bird *The Motor Car, 1765–1914* (1965)
The Motor-Car Index, 1918–29 (reprinted 1961)
The Motor-Car Index, 1928–39 (reprinted 1964)
Commercial Motor Index, 1913–24 (reprinted 1964)
C. Clutton and J. Stanford *The Vintage Car* (1954)
C. Clutton, C. Posthumus and D. Jenkinson *The Racing Car* (1956)
E. L. Cornwell *Commercial Road Vehicles* (1960)
R. Crabb *Birth of a Giant* (1969)
A. C. Harmsworth (ed.) *Motors and Motor Driving* (Badminton Library, 1902)
Institution of Automobile Engineers: Proceedings (from 1906–1907)
L. P. Jarman and R. Barraclough *The Bullnose Morris* (1965)
C. Jarrott *Ten Years of Motors and Motor-Racing* (1906)
K. Karslake and L. Pomeroy, Junior *From Veteran to Vintage* (1956)
P. W. Kingsford *F. W. Lanchester: the Life of an Engineer* (1960)
H. McCausland *The English Carriage (1948)*
J. McLellan *Bodies Beautiful* (1975)
Lord Montagu *Lost Causes of Motoring* (1960)
C. W. Morton *A History of Rolls-Royce Motor Cars, 1903–1907* (1964)
L. Pomeroy, Junior *The Grand Prix Car* (1954)
L. Pomeroy, Junior *Historic Racing Cars, 1907–60* (1963)
A. Harding (ed.) *Car Profiles* (1966–67/1972–73)
A. W. Soutter *The American Rolls-Royce* (1976)
D. Scott-Moncrieff *The Veteran and Edwardian Motor-Car* (1955)
J. Stanford *The Sports Car* (1962)
M. Sedgwick *Cars of the Thirties* (1970)
G. Sturt *The Wheelwright's Shop* (1923)
G. Turner *The Car Makers* (1964)
R. C. Wheatley and B. Morgan *The Restoration of Vintage & Thoroughbred Cars* (1957)
F. Young *The Complete Motorist* (1904)

Index

NOTE: the figures in bold refer to plate numbers and those in roman to page numbers.

Acknowledgements

The author and publishers would like to record their appreciation for the help received from the following individuals and organisations for providing the drawings and photographs listed below, and to thank them all for their permission to reproduce these. Any material that has not been acknowledged here is in the possession of the author.

Aston-Martin Ltd, 238; The Austin Motor Co, *219*; British Motor Corporation, *224*; British Leyland, *260*, *266*, *267*, *268*, *271*, *275*, *275a*, *275b*, *275c*, *281*, *282*, *284*; The Bristol Aeroplane Co, *239*; Carrozzeria Pininfarina, *254*, *262*; Citroën Cars Ltd, *150*, *256*, *264*; The Daimler Co, *247*; Doune Motor Museum, *114*, *158*, *159*, *161*, *164*, *188*, *205*; ERF Ltd, *252*; Esso Petroleum Co, *49*, *78*, *89*, *173*; FIAT, *23*, *55*, *269*; Ford Motor Co, *33*, *179*, *198*, *202*, *251*, *272*, *276*, *283*, *285*; General Motors, *212*; Jack Halley, *39*, *42*; IBCAM Library, *2*, *3*, *4*, *5*, *7*, *12*, *14*, *15*, *18*, *19*, *21*, *22*, *24*, *25*, *26*, *29*, *30*, *32*, *65*, *66*, *108*, *214* (Col. pls VIII, IX, X); Leslie Ireland, *101*, *106*, *107*, *111*, *129*, *148*, *217*, *218*, *221*, *223*, *273*, *274*; Imperial War Museum, *73*; Jensen Motors Ltd, *237*; Jaguar Cars Ltd, *216*; Lagonda Ltd, *232*; Leyland Vehicles, *250*, *253*; *Motor* magazine (Col. pls XIII–XVII); National Motor Museum, *84*, *85*, *91a*, *91b*, *93*, *96*, *103*, *104*, *144*, *146*, *151*, *157a*, *194a*, *194b*; Norco, *251*; The Nuffield Organisation, *220*, *240*, *270*; H. R. Owen Ltd, *243*, *244*; Renault Ltd, *265*; Rolls-Royce Motors, *227*, *234*, *235*, *245*, *246*; The Rover Co, *215*, *259*; Saab (Great Britain) Ltd, *277*; Standard-Triumph, *241*; Vauxhall Motors Ltd, *168*, *210*, *211*, *212*, *279* (Col. pl. XXV); Ab. Volvo, *278*; Worthington-Williams Collection, *99*, *165*, *201*, *286* (Col. pls XIII–XVII); Volkswagen Motors Ltd, *263*.